Nano-Technological Intervention in Agricultural Productivity

Nano-Technological Intervention in Agricultural Productivity

Javid A. Parray
Department of Higher Education
Government Degree College Eidgah
Srinagar, India

Mohammad Yaseen Mir
Department of School Education
University of Kashmir
Srinagar, India

Nowsheen Shameem
Department of Environmental Science
Cluster University Srinagar
Jammu and Kashmir, India

WILEY

This edition first published 2021

Registered Office
John Wiley & Sons, Inc., 111 River Street, Hoboken, NJ 07030, USA
John Wiley & Sons Ltd, The Atrium, Southern Gate, Chichester, West Sussex, PO19 8SQ, UK

Editorial Office
The Atrium, Southern Gate, Chichester, West Sussex, PO19 8SQ, UK

For details of our global editorial offices, customer services, and more information about Wiley products visit us at www.wiley.com.

Wiley also publishes its books in a variety of electronic formats and by print-on-demand. Some content that appears in standard print versions of this book may not be available in other formats.

Library of Congress Cataloging-in-Publication Data applied for

Hardback ISBN: 9781119714859

Cover Design: Wiley
Cover Image: © xuanhuongho/Shutterstock

Set in 9.5/12.5pt STIXTwoText by Straive, Chennai, India
Printed and bound by CPI Group (UK) Ltd, Croydon, CR0 4YY

C9781119714859_300621

Contents

About the Authors

Dr. Javid A. Parray, PhD, is an Assistant Professor in the Higher Education Department at the Government Degree College Eidgah, Srinagar, India, where he teaches the subject of environmental science. He has published many research articles in reputable refereed international and national journals. He had authored books on different themes, including Approaches to Heavy Metal Tolerance in Plants; Sustainable Agriculture: Biotechniques in Plant Biology; and Sustainable Agriculture: Advances in Plant Metabolome and Microbiome, soil bioremediation, climate change and microbes, etc. He was awarded the 'Scientist of the Year Award' by the Indian Academy of Environmental Science for the year 2015 in addition to being awarded with an international travel grant for participating in international conferences. He is currently the editor and reviewer of many reputed journals. He is also the founder of Academy of Ecoscience. Dr. Parray completed a fast-track project entitled 'Molecular characterization and metabolic potential of rhizospheric bacteria from *Arnebia benthamii* across North Western Himalaya' at CORD, University of Kashmir, India. He finished his postdoctoral research associateship on a DBT-funded project entitled 'Tissue culture-based network programme on saffron'. He earned a PhD degree in environmental science with a specialization in plant microbe interactions on the topic 'Evaluating role of rhizospheric bacteria in saffron culture' from the University of Kashmir, India. Dr. Parray is also the subject course coordinator for Environmental Science Approved CEC-Moocs by MHRD, Government of India, New Delhi.

Dr. Mohammad Yaseen Mir was working as a senior researcher in Research project funded by Ministry of Environment, Forest & Climate Change (MoEF & CC), Government of India along with G.B. Pant National Institute of Himalayan Environment & Sustainable Development under National Mission on Himalayan Studies (NMHS) and has recently joined Department of Education J&K as a teacher. He did his M.Phil and Ph.D Programmes from the University of Kashmir. He published many research articles in reputed, referred international and national journals like Springer/Elsevier/Hindwaii. He has also authored book Sustainable Agriculture: Biotechniques in Plant Biology with springer.

Dr. Nowsheen Shameem (M.Phil, Ph.D) is currently working as an Assistant Professor (CL) in the Department of Environmental Science at Cluster University Srinagar, Jammu and Kashmir, India. She did her doctorate (PhD) from the University of Kashmir on nutraceutical value of wild mushrooms growing in Kashmir valley in the year 2017. She has research expertise in microbial biotechnology, plant bioactivity and biotechnology, medicinal chemistry, and plant microbe interactions. She published many research articles in reputed, referred international and national journals such as Springer, Elsevier, and Hindawi. She has also authored many books on themes such as fresh water microbiology, plant biotechniques, advances in plant metabolome and microbiome, and climate changes and microbes. She is also the recipient of the Junior Scientist of the Year award by the International Foundation for Environment & Ecology.

About the Book

Contemporary agriculture shares the important portion of global economy for any sought-after growth, especially for its major contribution to human upliftment through poverty eradication, fast industrialization, financial change and investment, sustainable resource usage, and environmental management. The miniature aspect of nanotechnology controls major agricultural processes because of its diminutive dimensions. In addition, the use of nanotechnologies can be resonant obstruction, thanks to the many potential advantages such as enhancing the quality and safety of food, decreasing agricultural inputs, and enhancing soil absorption of nanoscale nutrients. This book will be immensely helpful to the students of plant biotechnology, agricultural sciences, and agricultural engineering students of both undergraduate and postgraduate levels in universities, colleges, and research institutes. The book will support researchers who work in the field of plant biotechnology and agricultural sciences. It is hoped this book will be another step towards the beneficial approach in plant biotechnology and setting of a new arena in shaping the new biotechniques towards the sustainable cause.

Key features:

1. Nanotechnological innovations in plant biology
2. Nanotechnology and transgenic via genome editing towards sustainable agricultural systems
3. Nanofertilizers and nanopesticides
4. Nanoparticle protection in plants.

1

Nanotechnology and Nanoparticles

CHAPTER MENU

1.1 Nanoparticles and Their Functions

Since the past century, nanotechnology has been an advanced research area. RP Feynman coined the term 'nanotechnology' during his famous speech [1]. Nanotechnology developed different types of nanoscale materials. A limited class of one-dimensional materials <100 nm [2] are nanoparticles (NPs). Depending on the form [3], such materials may be 0D, 1D, 2D, or 3D. The importance of these materials showed that the physicochemical

Nano-Technological Intervention in Agricultural Productivity, First Edition.
Javid A. Parray, Mohammad Yaseen Mir and Nowsheen Shameem.

properties of a substance, e.g. the optical properties, can be influenced by size. The red wine, yellowish-grey, black, and deep green are the distinctive colours of the 20 nm gold (Au), platinum (Pt), silver (Ag), and palladium (Pd) NPs. These NPs exhibit distinctive colours and properties of various sizes and shapes that can be used in bioimaging applications. [4]. Owing to differences in aspect ratio, nanoshell thickness, and Au concentration, the solution's colour varies. Changing each of the elements mentioned above affects the absorption characteristics of the NPs and is therefore observed in different absorption colours. Usually, NPs consist of three layers: (i) a layer-functionalized surface with several tiny molecules, metal ions, surfactants, and polymers; (ii) a shell layer – a chemically separate core substance; and (iii) a centre – an integral part of the NP and typically refers to the NP itself [5]. Researchers gained immense interest in multidisciplinary fields because of these exceptional features. NPs may be used for the delivery of drugs [6], chemical and biological sensing [7], gas sensing [8], CO_2 capture [9], and related uses [10].

1.2 Classification of NPs

NPs are commonly classified according to morphology, size, and chemical properties in different classes. NPs are classified according to physical and chemical characteristics as follows.

1.2.1 Carbon-Based NPs

Two key NP groups based on carbon are fullerenes and carbon nanotubes (CNTs). Fullerenes contain globular hollow cage nanomaterials, similar to allotropic carbon forms. Their electrical conductivity, high power, structure, electron affinity, and flexibility have created considerable commercial interest [11]. Pentagonal and hexagonal C- units were arranged in these materials while each carbon was hybridized. The illustration in Figure 1.1 shows some of the well-known 7.114 and 7.648 nm C_{60} and C_{70} fullerenes. The long, tubular CNTs have a diameter of 1–2 nm [12]. This can be predicted as metallic or semiconducting based on their telicity diameter [13]. Single, double, or multiple walls may be rolling sheets labelled single-walled nanotubes (SWNTs), double-walled nanotubes (DWNTs), or multi-walled carbon nanotubes (MWNTs), respectively. They are commonly synthesized by carbon precursor deposition, particularly atomic carbons, which are vaporized by a laser or electric arc through metal particles. Recently, they have been synthesized using the chemical vapour deposition (CVD) technique [14]. Because of their specific physical, chemical, and mechanical characteristics, these materials are widely used in industrial applications, not only in pristine form but also in nanocomposites such as fillers [15], active gas adsorbents for the remediation of the environment [16], and, in general, for various inorganic and organic catalysts [17].

1.2.2 Metal Nanoparticles

Metal NPs comprise metal-made precursors. Because of the well-localized surface plasma resonance (LSPR) characteristics, such NPs have distinct optoelectronic properties. In the visible region of the solar electromagnetic spectrum, the NPs of alkali and noble metals

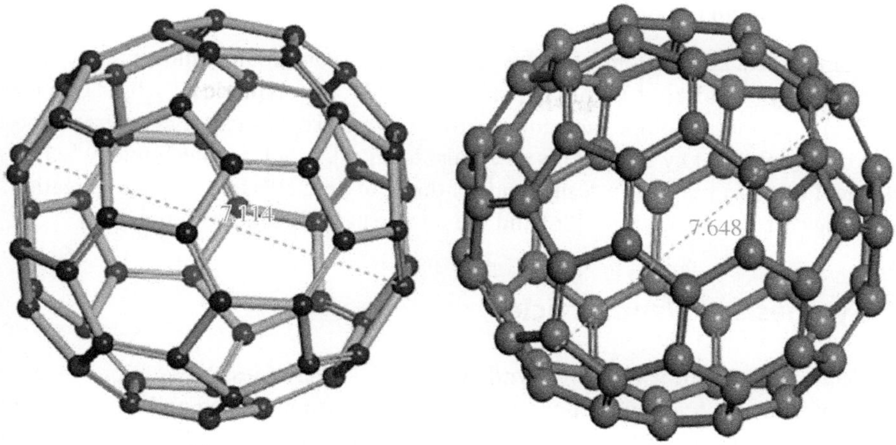

Figure 1.1 Description of fullerenes or buck balls (a) C_{60} and (b) C_{70}. Source: From Khan et al. [2]. © 2017, Elsevier.

such as Cu, Ag, and Au have broadband absorption. In today's cutting-edge materials, the facet, size, and shape-controlled synthesis of metal NPs are critical [4]. Metal NPs are finding applications in many research fields because of their advanced optical properties. The coating of gold NPs is widely used to enhance electronic streaming for scanning electron microscopy (SEM) sampling, thus helping to accomplish good SEM images.

1.2.3 Ceramic NPs

NPs from ceramics are inorganic non-metallic solids synthesized by heat and cooling. They are used in amorphous, polycrystalline, solid, porous, or hollow [18] forms. Because of their use in catalysis, photocatalysis, colour photodegradation, and imaging, these NPs offer substantial interest from researchers [19].

1.2.4 Semiconductor NPs

Semiconductor materials have properties between metals and non-metals, and because of this property, they find different applications in the literature [20]. Semiconductor NPs have large band gaps, demonstrating significant modifications to their properties with band gap tuning. Items of great significance also include photocatalysis, photo-optics, and electronic devices [21]. Because of their appropriate band gap and band edge positions, the range of semiconductor NPs is considered exceptionally efficient in water splitting applications [22].

1.2.5 Polymeric NPs

These substances are used for a wide range of commercial applications, such as fillers [15], effective adsorbents of environmental remedial gases [16], and as a support medium for various inorganic and organic calculators because they give specific physical, chemical, and mechanical characteristics [23].

1.2.6 NPs Based on Lipids

In many biomedical applications, they are used, and they contain lipid molecules. Usually, the lipid NP is spherical, 10–1000 nm in diameter. Similar to polymeric NPs, lipid NPs have a solid lipid core and lipophilic molecules within the matrix. The external centre of these NPs has been stabilized by surfactants or emulsifiers [24]. Lipid nanotechnology [25] is a specific field that focuses on the design and synthesis of lipid NPs for various applications, such as drug carriers and delivery [26] and the release of RNA cancer therapy [27].

1.3 Synthesis of Nanoparticles

Various techniques can be used for the synthesis of NPs. Nevertheless, these approaches are generally divided into two main categories, i.e. (i) top-down approach and (ii) bottom-up approach [28, 29] (Figure 1.2). These methods are further divided into different subclasses based on process, reaction state, and adopted protocols.

1.3.1 Top-Down Synthesis

The larger molecules are disintegrated into smaller units by a destructive process, and these units are transformed into useful NPs, for example, grinding/milling, CVDs, physical vapour deposition, etc. [29]. This approach is generally used to synthesize NPs from coconut shells (CSs). The milling method is used, whereby raw CS powders were finely milled using ceramic balls and a well-known planetary mill at different time intervals. Via other characterization techniques, the influence of the milling period on the overall size of the NPs is shown.

Furthermore, as time increases, the size of the NP crystallite (Scherrer equation) decreases. In this process, it was also found that the brownish colour faded away with the increment of each hour because of the reduced size of the NPs [30]. Various

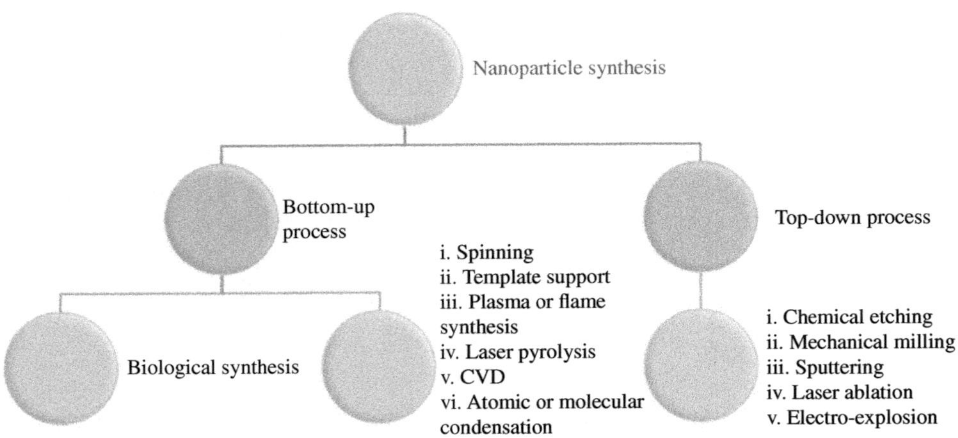

Figure 1.2 The synthetic models of NPs: top-down and bottom-up approach. Source: Modified from Iravani [29].

characterization techniques demonstrated the effect of milling time on the overall size of the NPs. The synthesis of spherical magnetite NPs from natural iron oxide (Fe_2O_3) ore was shown in the presence of organic oleic acid by a destructive top-down method with a particle size ranging from 20 to 50 nm [31]. To synthesize spherical particles of colloidal carbon using a control scale, a primary top-down route was used. The synthesis technique was based on the continuous chemical adsorption of polyoxometalates on the carbon interfacial surface. Adsorption has transformed black carbon aggregates into relatively smaller spherical particles with a high dispersion capacity and a narrow distribution of size [32]. Microstructures have found that the quantity of carbon particles is lower during the sonication period. Combined grinding and top-down sonication techniques synthesized a sequence of transition metal dichalcogenide nanodots (TMD-NDs) from their crystallites. Every TMD-ND with a size of less than 10 nm shows excellent dispersion and is demonstrated by the narrow distribution of the measure [33]. Highly photoactive Co_3O_4 NPs have recently been produced by top-down laser fragmentation, i.e. a top-down process with an average size of 5.8 ± 1.1 nm. Powerful laser irradiations produce well-uniformed NPs with adequate oxygen vacancy [34].

1.3.2 Bottom-Up Synthesis

This reverse approach is used to synthesize NPs from relatively more straightforward substances and is also called an approach to building up. Examples include sedimentation and reduction techniques. It encompasses sun-freezing, green synthesis, spinning, and biochemical synthesis [29]. Mogilevsky et al. [35] used this technique to synthesize TiO_2 anatase NPs to the graphene domains. Alizarin and titanium isopropoxide precursors have been used to synthesize the photoactive composite for methylene blue photocatalytic degradation. The X-ray diffraction (XRD) framework has verified the anatase form [35]. Well-uniform spherical shaped Au nanospheres have been synthesized with monocrystalline using top-down laser irradiation technique [36, 37]. Recently, the solvent exchange method has been used to deliver medical cancer drugs to achieve limited-size low-density lipoprotein (LDL) NPs. The standard approach followed by growth, i.e. up process, is nucleation in this technique. The LDL NPs were obtained without phospholipid use and had high hydrophobicity, which is a prerequisite for drug delivery implementation. [38]. The monodispersed spherical bismuth (Bi) NPs, with top-down and bottom-up approaches, are synthesized with excellent colloidal properties [39]. In bottom-up ethylene glycol, bismuth acetate was melted, although bismuth was converted into a molten state in the top-down process. In boiled diethylene glycol, the molten drop then has been emulsified for NPs. Both methods generated different NPs in size from 100 to 500 nm [39] (Figure 1.3a,b). Green and biogenic bottom-up processing is cost-effective and environmentally sustainable, where biological processes, such as using plant extracts, achieve the synthesis of NPs. For the synthesis of NPs, bacteria, yeast, fungi, *Aloe vera*, tamarind, and even human cells are used. Au-NPs were synthesized from wheat biomass and oat [40] and used as a reduction agent by microorganisms and plant extracts [41, 42]. Figure 1.3 demonstrated the bottom-up (Figure 1.3a) method: decomposing a molecular precursor into simple metal atoms, which transform into colloids and the top-down (Figure 1.3b) method.

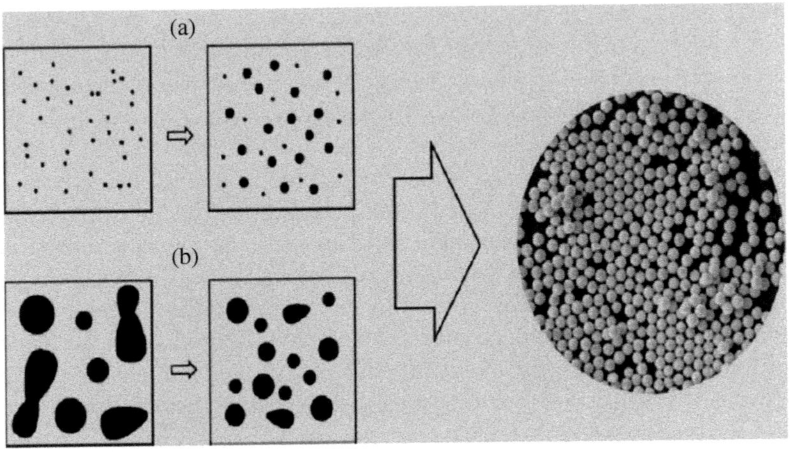

Figure 1.3 Illustration of synthesis of nanoparticles: (a) top-down method and (b) bottom-up method. Source: From Wang and Xia [39]. © 2004, American Chemical Society.

1.4 NPs and Characterization

For the analysis of other physicochemical properties of NPs, different methods of characterization have been used, such as XRD, X-ray photoelectron spectroscopy (XPS), infrared (IR), SEM, transmission electron microscopy (TEM), and Brunauer–Emmett–Teller (BET). Advanced methods are applied for the analysis of the particles.

1.4.1 Morphological Characterization

The morphological characteristics of NPs are still of great importance as morphology still affects most of the NP properties. Various characterization techniques exist for morphological research, but microscopic methods exist, such as polarized optical microscopy (POM), SEM, and TEM.

1.4.1.1 SEM Technique

The SEM technique is based on the electron scanning principle and provides all the nanoscale NP data available. This technique is used to study the morphology of their nanomaterials and the dispersion of NPs in the bulk or matrix. This technique [15] showed the distribution of SWNTs in polymer matrix poly(butylene) terephthalate (PBT) and nylon-6. The morphological characteristics of ZnO-modified metal–organic frameworks (MOFs) were studied using the SEM technique, which indicates the dispersion of ZnO-NPs and MOFs' morphologies under different reaction conditions [43].

1.4.1.2 TEM Technique

It is based on the electron transfer principle, so that it can provide descriptions of the bulk content from very low to greater magnification. In addition, it is commonly used for the analysis of different morphologies of the Au-NPs [44]. TEM also provides essential information about two- or more layer materials; the quadruple hollow shell structure of Co_3O_4

NPs is observed by TEM, for instance. In Li-ion batteries such as the anode, these NPs have proven themselves to be exceptionally efficient. The porous multi-shell structure induces shorter Li^+ diffusion path lengths with ample annulled space for buffer volume expansion, good cycling efficiency, higher speed capacity, and essential capacity [45].

1.4.2 Structural Characteristics

The structural characteristics of the structure and function of the bonding materials are of primary importance for studying. It gives details about the bulk properties of the subject material. XRD, energy-dispersive X-ray (EDX) spectroscopy, XPS, IR, Raman spectroscopy, BET, and Zieta size analyser are the techniques used to study the structural properties of NPs.

1.4.2.1 XRD

One of the essential characterization techniques is to reveal the structural properties of NPs. The crystalline phase of NPs is provided with sufficient data. It also provides a rough image of the particle size through the Debye–Scherrer [8] formula. In the identification of single and multiphase NP [46] schemes, this approach worked well. However, in smaller NPs with a size smaller than hundreds of atoms, the acquisition and accurate measurement of structural and other parameters may be difficult. Besides, the XRD diffractogram can be affected by NPs with different interatomic lengths having more amorphous characteristics. To obtain accurate data, the diffractograms of bimetallic NPs must be contrasted with those of the corresponding monometallic NPs and their physical mixtures in this case. The best way to make a substantial difference is to measure the simulated bimetallic NP structural model with the spectra of XRD [47] observed.

1.4.2.2 Energy-Dispersive X-ray (EDX)

To understand the elementary composition with a rough idea of per cent weight, a usually fixed field emission scanning electron microscopy (FE-SEM) or TEM system is commonly used. The electron beam centred on a single NP through SEM or TEM through the software functions to obtain the insight knowledge under observation from the NP. NP consists of constituent elements and, by irradiating electron beams, each of these releases X-ray energy characteristics. The real X-ray intensity is directly proportional to the explicit part of the particle's concentration. Researchers in preparatory materials commonly use this technique to help SEM and other processes to validate their components [48]. The elemental composition of ultra-sonochemically synthesized $BiVO_4$ NPs in pseudo-flower form [49] was calculated using the EDX technique. Similarly, a similar approach was used to perform the indispensable confirmation and graphene impregnation of Ln_2O_3/graphene heterostructure NPs, which showed C, Ln, and O as contributing elements synthesized by the traditional hydrothermal method [50].

1.4.2.3 XPS

It is a surface-sensitive tool and can be used to consider the overall composition and the compositional variance with in-depth profiling studies. XPS is based on the basic principles of spectroscopy. The typical XPS spectrum consists of the number of electrons on the Y-axis plot versus the X-axis electrons' binding energy (eV). Each element has its fingerprint value for energy binding and thus gives a particular set of XPS peaks. Corresponding peaks, such as 1s, 2s, 2p, and 3s, come from the electronic configuration [51]. To research the dispersion of Boron NPs (10 nm size) during functionalization with polyethylene glycol (PEG), a depth profile analysis was given with Ar^+ ions at 1.4 keV and 20 nm. It has been shown that the concentration of NPs increases from 2% to 5% with depth. This offered strong evidence that within the bulk of functionalized PEG, boron NPs are effectively dissolved [52]. In a related analysis, core–shell Au/Ag showed similar behaviour through XPS scope profiling [53].

1.4.2.4 FT-IR and Raman Spectroscopies

The vibration characterization of NPs is typically studied by FT-IR and Raman spectroscopies. These techniques are the most evolved and feasible compared to other simple analytical methods. The critical range for NPs is the fingerprint region, which provides the details for the material signature. In one sample, Pt-NP (1.7 nm) functionalization and its interaction with the alumina substrate were analysed using FT-IR and XPS techniques. FT-IR confirms the functionality as it showed signature vibrational peaks of carboxylated C–O 2033 cm^{-1}, in addition to a broader O–H peak of 3280 cm^{-1}, respectively [54, 55]. Because of its signal-enhancing capability via SPR phenomenon, recently improved surface-enhanced Raman spectroscopy (SERS) is emerging as a vibrational conforming tool [56, 57].

1.4.3 Particle Size and Surface Area Characterization

Various techniques can calculate the size and surface area of the NPs. These include dispersing SEM, TEM, XRD, AFM, and dynamic light scattering (DLS). It is possible to increase the particle size of SEM, TEM, XRD, and AFM, but the zeta theoretical analyser/DLS should be used to find the NP size at a weak stage. In one study, DLS was used to analyse silica NP size differences while consuming serum proteins. With the acquisition of the protein layer, the findings showed that the size increased. However, in the case of aggregation and hydrophilicity, DLS might prove incapable of accurate measurement, so in that case, we should focus on the high-resolution technique of differential centrifugal sedimentation [58–60].

1.4.4 Optical Characterizations

Optical characteristics are of great concern in photocatalytic applications, and photochemists have therefore gathered a good understanding of this approach to reveal their photochemical processes [61, 62]. These are based on the common law of Beer–Lambert and the basic principles of light [63]. These methods include information about the absorption, reflectance, luminescence, and phosphorescence of nanomaterials. Metallic and semiconductor NPs have various colours and are thus ideally suited for photo-related applications. To understand each application's primary mechanism, it is often essential to see the importance of absorption and reflectance of these materials. UV–vis and photoluminescence are the most common optical devices used to study the optical properties of NP materials (PL, Null Ellipsometer). The Diffuse reflectance spectroscopy (DRS) UV/vis is a fully designed optical absorption, transmission, and reflection measurement unit. The first two are extra, while DRS is mostly a unique technique for the samples sold. It is imperative to use the method for measuring NP band gaps as well as other NPs. MMT, $LaFeO_3$, and $LaFeO_3$/MMT nanocomposite synthesis and differences in their absorption of electromagnetic radiation by UV–vis DRS to identify their optical characteristics were studied [64]. In the case of nanocomposites, a significant red shift was observed compared to pristine MMT and $LaFeO_3$ NPs. Instead of a broad absorption band from 400 to 620 nm, $LaFeO_3$ and $LaFeO_3$/MMT revealed a reduction in their band gap. These NPs are significant for photocatalysis by solar light [64]. To investigate the optical properties of photoactive NPs and other nanomaterials and UV, PL considers useful technologies. This technique gives further information on the absorption or emission potential of the materials and their effects on the picture's overall excitement period. It thus provides valuable details about the charging hybridization and half-life of the exciting material on their conducting bands for all photo- and image applications.

1.5 Physicochemical Properties of NPs

Different physicochemical features such as the large surface are discussed; as previously mentioned, mechanically robust, optically active, and chemically reactive NPs are unique and ideal for multiple uses. Some of its essential properties are discussed in the following.

1.5.1 Mechanical and Optical Properties

There is greater interdependence between the optical and electronic properties of NPs. The noble metal NPs, for example, display full UV–visible extinction bands not available on the bulk metal spectrum and have visual properties that are dependent on size. When the conduction electrons' mutual excitation is aroused, this band of enthusiasm results in a continuous photon occurrence, known as the LSPR. LSPR excitation results in wavelength selection absorption with a large Ray light scattering coefficient of molar excitation resonance with an efficiency equal to that of 10 fluorophores and enhanced local electromagnetic fields near the surface of NPs, which strengthened spectroscopy. It is well known that the absorption spectrum of the LSPR spectrum relies on the size, shape, and interparticle spacing of the NPs, as well as its dielectric and local characteristics, such as substrates, solvents, and adsorbents [65]. The rusty colours seen in the door/windows of blemished glass are gold colloidal NPs responsibility, while Ag NPs are usually yellow. The free electrons on the surface are easily transportable via the nanomaterial in these NPs (d electrons in Ag and gold). For Ag and gold, the mean open path is 50 nm, more than the size of these materials in NPs. Thus, no scattering is required from the bulk after weak interaction. Instead, in these NPs, they set up standing resonance conditions responsible for LSPR [66, 67].

1.5.2 Magnetic Properties

For researchers from various disciplines, including heterogeneous and homogeneous catalysis, biomedicine, magnetic fluids, data storage for magnetic resonance imaging (MRI), and environmental remediation such as water decontamination, magnetic NPs are of considerable interest. The literature indicates that NPs work better when the size is smaller than the critical value, i.e. 10–20 nm [68]. Effectively controlled at such a low scale, the magnetic properties of NPs make these particles priceless and can be used in different applications [31, 68, 69]. In NPs, the uneven electronic distribution causes the magnetic property development [70, 71].

1.5.3 Mechanical Properties

Researchers can find new applications in a wide range of necessary sectors, including tribology, surface engineering, and nano-making, thanks to its distinct mechanical properties. A mechanical study of the automated nature of the NPs involves elastic modulus, hardness, stress, vibration, adhesion, and friction. Coagulation and lubrication also help improve the mechanical characteristics of the NPs and this parameter [72]. NPs exhibit different mechanical properties in contrast with microparticles and their bulk materials. Furthermore, comparing the steepness between NPs and the external contact surface checking on a lubricated or grated contact reveals that the NPs operate in a communication setup. Decent checks and interactions between the NPs' mechanical characteristics and any surface shape are critical for improving surface quality and enhancing material elimination. In these areas, a strong understanding of the fundamental mechanical aspects of NPs, including the elastic module and the hardness, motion, friction, and input, typically requires good performance [72].

1.5.4 Thermal Properties

The thermal conductivity of NP metals is known to be higher than that of stable fluids. For example, the thermal copper conductivity is about 700 times higher than water and approximately 3000 times higher than engine oil at room temperature. In addition, alumina oxides (Al_2O_3) are thermally more thermally capable than water. Fluids containing solids suspended with higher thermal conductance should therefore be substantially higher than conventional heat transmission fluid. Dispersing the nanometric scales solid particles into liquid such as water, ethylene glycol or oils produces nanofluids. Dispersed nanometric scale nanofluids are supposed to exhibit superior propensities compared to conventional heat transfer fluids and fluids containing microscopic particles. As this thermal transfer occurs on the particles' surface, it is essential to use particles with a large overall surface region. The wider total area also improves the stability of the suspension [73]. It has recently been shown that advanced thermal conductivity is exhibited by nanofluids consisting of CuO or Al_2O_3 NPs in water or ethylene [74].

1.6 Functions of NPs

The NPs find their application in almost every day-to-day utility, and some of the significant applications are discussed as follows:

1.6.1 Drugs and Medications

The basic physical and chemical properties of single or multiple nano-sized inorganic particles are becoming an increasingly valuable commodity in developing novel nanoequipment used in many different physical, biological, biomedical, and therapeutic products [75].

Throughout every medicine market, the importance of NPs in delivering drugs in the best dosage range has improved. This has sometimes resulted in an improvement in the medicines' clinical efficiency, weakened side effects, and improved patient compliance [76]. Iron oxide particles such as magnetite (Fe_3O_4) or its oxidized form of maghemite (Fe_2O_3) are most widely used for biomedical applications [77]. For biological and cell imaging applications and photothermal therapeutic applications, the option of NPs to achieve efficient contrast is based on the optical properties of NPs [78]. Over the past few years, hydrophilic NP development as a drug carrier has represented a significant challenge. Polyethylene oxide (PEO) and polylactic acid (PLA) NPs were established as an excellent method for intravenous drug administration among the different approaches [79]. For various in vivo applications, such as MRI contrast enhancement, tissue repair and immunoassay, detoxification of biological fluids, hyperthermia, drug delivery, and cell separation, superparamagnetic iron oxide NPs with good surface chemistry can be used [80]. Antibodies labelled with fluorescent dyes, enzymes, radioactive compounds, or colloidal Au [67] can be used to detect analytes in tissue parts via antigen–antibody interactions.

The specific benefits of liposomes, including the capability to shield drugs from degradation, target them at their active sites, and minimize harmfulness and other side effects, are a possible carrier instead of traditional dosage types. The polymeric NPs provide some significant advantages over the liposomes of these materials. NPs, for example, help improve drug ratability and provide convenient, controlled drug release properties. The extreme absorbed light is effectively converted into localized heat by Au-NPs, which can be used for targeted laser photothermal cancer therapy [81, 82]. Besides this, to prevent tumour development, the antineoplastic effect of NPs is also effectively used. Compared to organic compounds that are comparatively toxic to biological systems, the antimicrobial properties of inorganic NPs add more potency to this essential feature [83–85]. To selectively overcome the microbial cells, the NPs are engaged with various classes. Because of their adequate antibacterial efficacy, TiO_2, ZnO, $BiVO_4$, and Cu- and Ni-based NPs have been used for this reason [86].

1.6.2 Materials and Manufacturing

The NP manufacturer displays physicochemical characteristics that induce specific electrical, mechanical, optical, and imaging properties that are highly sought for applications in the medical, commercial, and ecological sectors in particular [87]. NPs focus on biological and non-biological characterization, design, and engineering of <100 nm structures, exhibiting new and unique characteristics. Many producers at high and low levels have reported the potential benefits of nanotechnology. Mass production of marketable products such as microelectronics, aerospace, and pharmaceutical companies are now underway [88]. Many industries have indicated that nanotechnology is the next development, including food processing and packaging. RET (organic dye molecules and noble metals) has been considered in the recent interest in biophotonics and materials science [89]. The plasmon resonance arising from the reciprocal oscillation of electrons at the surface of the NP [89, 90] is distinctly coloured in NP metals, such as noble metals, such as Au and Ag.

1.6.3 Environment

The release of these materials to the atmosphere contributes to commercial and domestic engineered nuclear power plants [91]. The use of engineering materials would increase soil and groundwater NP concentrations, which provide the most significant exposure pathways for assessing environmental risk [92]. During the formation of natural NPs, the surface of NPs can be consumed, co-precipitated, or stuck with the accumulation of NPs containing toxins adsorbed to their bodies by a vast specific-to-mass proportion of natural NPs. NP pollutants' interaction depends on the characteristics of NPs, such as scale, composition, morphology, porosity and aggregation, and structure [93]. The following attributes of NPs make the ideal theme candidate for environmentally friendly goods, sanitation of toxic substance-contaminated materials, and ecological stage sensors [10]. Superparamagnetic iron oxide NPs are a valuable sorbent material for this harmful soft material [94, 95].

NP photodegradation is also a generalized method, which includes the use of several nanomaterials. For photodegradation, Rogozea et al. revealed in a tandem fashion that modified silica NiO/ZnO has been productive because of the minimum size of the high NP surface (<10 nm) [96].

1.6.4 Electronics

In recent years, there has been rising interest in printed electronics production because printed electronics offer the potential for low-cost, large-area electronics for flexible displays and sensors appealing to conventional silicon techniques. As a mass manufacturing process for new forms of electronic equipment, printed electronics with various functional inks containing NPs such as metallic NPs, organic electronic molecules, CNTs, and ceramic NPs are expected to flow quickly [97, 98]. An excellent example of the synergies between scientific discovery and technological growth is the electronic industry. The findings of new semiconducting materials have led to a revolution from aspirated tubes to diodes and transistors and finally to miniature chips [10, 99]. The critical characteristics of NPs that make nanotechnology benchmarks [100] possible for NP to be used in electrical, electronic, or optical applications, including bottom-up or self-assembly frameworks, are easy handling.

1.6.5 Energy Harvesting

Because of their large surface area, optical behaviour, and catalytic nature, scientists are changing their research strategies to produce renewable energies from readily available resources at low cost. NPs are the best candidate for this reason. NPs are widely used to generate power from photoelectrochemical (PEC) and electrochemical water splitting [48], especially in photocatalytic applications. Electrochemical CO_2 reduction in fuel precursors, solar cells, and piezoelectric generators also provided advanced energy generation options in addition to water splitting [34]. NPs are often used in energy storage applications to reserve animations in various ways at the nanoscale level [101, 102]. Nanogenerators have recently been developed to transform mechanical energy into electricity using piezoelectric power, an unconventional approach to power generation [103].

References

1 Feynman, R.P. (1960). There's plenty of room at the bottom. *Eng. Sci.* 22: 22–36.
2 Khan, I., Saeed, K., and Khan, I. (2017). Nanoparticles: properties, applications and toxicities. *Arabian J. Chem.* 12: 908–931.
3 Tiwari, J.N., Tiwari, R.N., and Kim, K.S. (2012). Zero-dimensional, one-dimensional, two-dimensional and three-dimensional nanostructured materials for advanced electrochemical energy devices. *Prog. Mater. Sci.* 57: 724–803. https://doi.org/10.1016/j.pmatsci.2011.08.003.
4 Dreaden, E.C., Alkilany, A.M., Huang, X. et al. (2012). The golden age: gold nanoparticles for biomedicine. *Chem. Soc. Rev.* 41: 2740–2779. https://doi.org/10.1039/C1CS15237H.
5 Shin, W.K., Cho, J., Kannan, A.G. et al. (2016). Cross-linked composite gel polymer electrolyte using mesoporous methacrylate-functionalized SiO_2 nanoparticles for lithium-ion polymer batteries. *Sci. Rep.* 6: 26332. https://doi.org/10.1038/srep26332.
6 Lee, J.E., Lee, N., Kim, T. et al. (2011). Multifunctional mesoporous silica nanocomposite nanoparticles for theranostic applications. *Acc. Chem. Res.* 44: 893–902. https://doi.org/10.1021/ar2000259.
7 Barrak, H., Saied, T., Chevallier, P. et al. (2016). Synthesis, characterization, and functionalization of ZnO nanoparticles by *N*-(trimethoxysilylpropyl) ethylenediamine tri acetic acid (TMSEDTA): an investigation of the interactions between phloroglucinol and ZnO@TMSEDTA. *Arabian J. Chem.* https://doi.org/10.1016/j.arabjc.2016.04.019.
8 Ullah, H., Khan, I., Yamani, Z.H., and Qurashi, A. (2017). Sonochemical-driven ultrafast facile synthesis of SnO_2 nanoparticles: growth mechanism structural electrical and hydrogen gas sensing properties. *Ultrason. Sonochem.* 34: 484–490. https://doi.org/10.1016/j.ultsonch.2016.06.025.
9 Ramacharyulu, P.V.R.K., Muhammad, R., Praveen Kumar, J. et al. (2015). Iron phthalocyanine modified mesoporous titania nanoparticles for photocatalytic activity and CO_2 capture applications. *Phys. Chem. Chem. Phys.* 17: 26456–26462. https://doi.org/10.1039/C5CP03576G.
10 Shaalan, M., Saleh, M., El-Mahdy, M., and El-Matbouli, M. (2016). Recent progress in applications of nanoparticles in fish medicine: a review. *Nanomed. Nanotechnol. Biol. Med.* 12: 701–710. https://doi.org/10.1016/j.nano.2015.11.005.
11 Astefanei, A., Núñez, O., and Galceran, M.T. (2015). Characterization and determination of fullerenes: a critical review. *Anal. Chim. Acta* 882: 1–21. https://doi.org/10.1016/j.aca.2015.03.025.
12 Ibrahim, K.S. (2013). Carbon nanotubes-properties and applications: a review. *Carbon Lett.* 14: 131–144. https://doi.org/10.5714/CL.2013.14.3.131.
13 Aqel, A., El-Nour, K.M.M.A., Ammar, R.A.A., and Al-Warthan, A. (2012). Carbon nanotubes, science and technology part (I) structure, synthesis and characterization. *Arabian J. Chem.* 5: 1–23. https://doi.org/10.1016/j.arabjc.2010.08.022.

14 Elliott, J.A., Shibuta, Y., Amara, H. et al. (2013). Atomistic modelling of CVD synthesis of carbon nanotubes and graphene. *Nanoscale* 5: 6662. https://doi.org/10.1039/c3nr01925j.

15 Saeed, K. and Khan, I. (2016). Preparation and characterization of single-walled carbon nanotube/nylon 6,6 nanocomposites. *Instrum Sci. Technol.* 44: 435–444. https://doi.org/10.1080/10739149.2015.1127256.

16 Ngoy, J.M., Wagner, N., Riboldi, L., and Bolland, O. (2014). A CO_2 capture technology using multi-walled carbon nanotubes with polyaspartamide surfactant. *Energy Procedia* 63: 2230–2248. https://doi.org/10.1016/j.egypro.2014.11.242.

17 Mabena, L.F., Sinha Ray, S., Mhlanga, S.D., and Coville, N.J. (2011). Nitrogen-doped carbon nanotubes as a metal catalyst support. *Appl. Nanosci.* 1: 67–77. https://doi.org/10.1007/s13204-011-0013-4.

18 Sigmund, W., Yuh, J., Park, H. et al. (2006). Processing and structure relationships in electrospinning of ceramic fiber systems. *J. Am. Ceram. Soc.* 89: 395–407. https://doi.org/10.1111/j.1551-2916.2005.00807.x.

19 Thomas, S., Harshita, B.S.P., Mishra, P., and Talegaonkar, S. (2015). Ceramic nanoparticles: fabrication methods and applications in drug delivery. *Curr. Pharm. Des.* 21: 6165–6188. https://doi.org/10.2174/1381612821666151027153246.

20 Ali, S., Khan, I., Khan, S.A. et al. (2017). Electrocatalytic performance of Ni@Pt core-shell nanoparticles supported on carbon nanotubes for methanol oxidation reaction. *J. Electroanal. Chem.* 795: 17–25. https://doi.org/10.1016/j.jelechem.2017.04.040.

21 Sun, S. (2000). Monodisperse FePt nanoparticles and ferromagnetic FePt nanocrystal superlattices. *Science* 80 (287): 1989–1992. https://doi.org/10.1126/science.287.5460.1989.

22 Hisatomi, T., Kubota, J., and Domen, K. (2014). Recent advances in semiconductors for photocatalytic and photoelectrochemical water splitting. *Chem. Soc. Rev.* 43: 7520–7535. https://doi.org/10.1039/C3CS60378D.

23 Abouelmagd, S.A., Meng, F., Kim, B.-K. et al. (2016). Tannic acid-mediated surface functionalization of polymeric nanoparticles. *ACS Biomater. Sci. Eng.*: 6b00497. https://doi.org/10.1021/acsbiomaterials.6b004.

24 Rawat, M.K., Jain, A., Singh, S. et al. (2011). Studies on binary lipid matrix-based solid lipid nanoparticles of repaglinide: in vitro and in vivo evaluation. *J. Pharm. Sci.* 100: 2366–2378. https://doi.org/10.1002/jps.22435.

25 Mashaghi, S., Jadidi, T., Koenderink, G., and Mashaghi, A. (2013). Lipid nanotechnology. *Int. J. Mol. Sci.* 14: 4242–4282. https://doi.org/10.3390/ijms14024242.

26 Puri, A., Loomis, K., Smith, B. et al. (2009). Lipid-based nanoparticles as pharmaceutical drug carriers: from concepts to clinic. *Crit. Rev. Ther. Drug Carrier Syst.* 26: 523–580.

27 Gujrati, M., Malamas, A., Shin, T. et al. (2014). Multifunctional cationic lipid-based nanoparticles facilitate endosomal escape and reduction-triggered cytosolic siRNA release. *Mol. Pharmaceutics* 11: 2734–2744. https://doi.org/10.1021/mp400787s.

28 Wang, Y. and Xia, Y. (2004). Bottom-up and top-down approaches to the synthesis of monodispersed spherical colloids of low melting-point metals. *Nano Lett.* 4: 2047–2050. https://doi.org/10.1021/nl048689j.

29 Iravani, S. (2011). Green synthesis of metal nanoparticles using plants. *Green Chem.* 13: 2638. https://doi.org/10.1039/c1gc15386b.

30 Bello, S.A., Agunsoye, J.O., and Hassan, S.B. (2015). Synthesis of coconut shell nanoparticles via a top-down approach: assessment of milling duration on the particle sizes and morphologies of coconut shell nanoparticles. *Mater. Lett.* https://doi.org/10.1016/j.matlet.2015.07.063.

31 Priyadarshana, G., Kottegoda, N., Senaratne, A. et al. (2015). Synthesis of magnetite nanoparticles by top-down approach from a high purity ore. *J. Nanomater.*: 1–8. https://doi.org/10.1155/2015/317312.

32 Garrigue, P., Delville, M.-H., Labruge're, C. et al. (2004). Top-down approach for the preparation of colloidal carbon nanoparticles. *Chem. Mater.* 16: 2984–2986. https://doi.org/10.1021/cm049685i.

33 Zhang, X., Lai, Z., Liu, Z. et al. (2015). A facile and universal top-down method for preparation of monodisperse transition-metal dichalcogenide nanodots. *Angew. Chem. Int. Ed.* 54: 5425–5428. https://doi.org/10.1002/anie.201501071.

34 Zhou, Y., Dong, C.K., Han, L. et al. (2016). Top-down preparation of active cobalt oxide catalyst. *ACS Catal.* 6: 6699–6703. https://doi.org/10.1021/acscatal.6b02416.

35 Mogilevsky, G., Hartman, O., Emmons, E.D. et al. (2014). Bottom-up synthesis of anatase nanoparticles with graphene domains. *ACS Appl. Mater. Interfaces* 6: 10638–10648. https://doi.org/10.1021/am502322y.

36 Liu, D., Li, C., Zhou, F. et al. (2015). Rapid synthesis of monodisperse Au nanospheres through a laser irradiation-induced shape conversion, self-assembly and their electromagnetic coupling SERS enhancement. *Sci. Rep.* 5: 7686. https://doi.org/10.1038/srep07686.

37 Liu, J., Liu, Y., Liu, N. et al. (2015). Metal-free efficient photocatalyst for stable visible water splitting via a two-electron pathway. *Science* 80 (347): 970–974. https://doi.org/10.1126/science.aaa3145.

38 Needham, D., Arslanagic, A., Glud, K. et al. (2016). Bottom-up design of nanoparticles for anti-cancer diapeutics: "put the drug in the cancer's food". *J. Drug Targeting*: 1–21. https://doi.org/10.1080/1061186X.2016.1238092.

39 Wang, Y. and Xia, Y. (2004). Bottom-up and top-down approaches to synthesizing monodispersed spherical colloids of low melting-point metals. *Nano Lett.* 4: 2047–2050. https://doi.org/10.1021/nl048689j.

40 Parveen, K., Banse, V., and Ledwani, L. (2016). Green synthesis of nanoparticles: their advantages and disadvantages. *Acta Naturae*: 20048. https://doi.org/10.1063/1.4945168.

41 Ahmed, S., Annu, S., and Yudha, S.S. (2016). Biosynthesis of gold nanoparticles: a green approach. *J. Photochem. Photobiol., B* 161: 141–153. https://doi.org/10.1016/j.jphotobiol.2016.04.034.

42 Biswas, A., Bayer, I.S., Biris, A.S. et al. (2012). Advances in top-down and bottom-up surface nanofabrication: techniques, applications and prospects. *Adv. Colloid Interface Sci.* 170: 2–27. https://doi.org/10.1016/j.cis.2011.11.001.

43 Mirzadeh, E. and Akhbari, K. (2016). Synthesis of nanomaterials with desirable morphologies from metal–organic frameworks for various applications. *CrystEngComm* 18: 7410–7424. https://doi.org/10.1039/C6CE01076H.

44 Khlebtsov, N. and Dykman, L. (2011). Biodistribution and toxicity of engineered gold nanoparticles: a review of in vitro and in vivo studies. *Chem. Soc. Rev.* 40: 1647–1671. https://doi.org/10.1039/C0CS00018C.

45 Wang, J., Yang, N., Tang, H. et al. (2013). Accurate control of multishelled Co_3O_4 hollow microspheres as high-performance anode materials in lithium-ion batteries. *Angew. Chem. Int. Ed.* 52: 6417–6420. https://doi.org/10.1002/anie.201301622.

46 Emery, A.A., Saal, J.E., Kirklin, S. et al. (2016). High-throughput computational screening of perovskites for thermochemical water splitting applications. *Chem. Mater.* 28 https://doi.org/10.1021/acs.chemmater.6b01182.

47 Ingham, B. (2015). X-ray scattering characterization of nanoparticles. *Crystallogr. Rev.* 21: 229–303. https://doi.org/10.1080/0889311X.2015.1024114.

48 Avasare, V., Zhang, Z., Avasare, D. et al. (2015). Room-temperature synthesis of TiO_2 nanospheres and their solar-driven photoelectrochemical hydrogen production. *Int. J. Energy Res.* 39: 1714–1719. https://doi.org/10.1002/er.3372.

49 Khan, I., Ali, S., Mansha, M., and Qurashi, A. (2017). Sonochemical assisted hydrothermal synthesis of pseudo-flower shaped Bismuth vanadate ($BiVO_4$) and their solar-driven water splitting application. *Ultrason. Sonochem.* 36: 386–392. https://doi.org/10.1016/j.ultsonch.2016.12.014.

50 Mansha, M., Qurashi, A., Ullah, N. et al. (2016). Synthesis of In_2O_3/graphene heterostructure and their hydrogen gas sensing properties. *Ceram. Int.* 42: 11490–11495. https://doi.org/10.1016/j.ceramint.2016.04.035.

51 Lykhach, Y., Kozlov, S.M., Skála, T. et al. (2015). Counting electrons on supported nanoparticles. *Nat. Mater.* https://doi.org/10.1038/nmat4500.

52 Oprea, B., Martínez, L., Román, E. et al. (2015). Dispersion and functionalization of nanoparticles synthesized by gas aggregation source: opening new routes toward the fabrication of nanoparticles for biomedicine. *Langmuir* 31: 13813–13820. https://doi.org/10.1021/acs.langmuir.5b03399.

53 Wang, Y.C., Engelhard, M.H., Baer, D.R., and Castner, D.G. (2016). Quantifying the impact of nanoparticle coatings and nonuniformities on XPS analysis: gold/silver core-shell nanoparticles. *Anal. Chem.* 88: 3917–3925. https://doi.org/10.1021/acs.analchem.6b00100.

54 Dablemont, C., Lang, P., Mangeney, C. et al. (2008). FTIR and XPS study of Pt nanoparticle functionalization and interaction with alumina. *Langmuir* 24: 5832–5841. https://doi.org/10.1021/la7028643.

55 Pokhrel, M., Wahid, K., and Mao, Y. (2016). Systematic studies on RE_2-Hf_2O_7:5%Eu^{3+} (RE = Y, La, Pr, Gd, Er, and Lu) nanoparticles: effects of the A-site RE^{3+} cation and calcination on structure and photoluminescence. *J. Phys. Chem. C* 120: 14828–14839. https://doi.org/10.1021/acs.jpcc.6b04798.

56 Muehlethaler, C., Leona, M., and Lombardi, J.R. (2016). Review of surface-enhanced Raman scattering applications in forensic science. *Anal. Chem.* 88: 152–169. https://doi.org/10.1021/acs.analchem.5b04131.

57 Ma, S., Livingstone, R., Zhao, B., and Lombardi, J.R. (2011). Enhanced Raman spectroscopy of nanostructured semiconductor phonon modes. *J. Phys. Chem. Lett.* 2: 671–674. https://doi.org/10.1021/jz2001562.

58 Sikora, A., Shard, A.G., and Minelli, C. (2016). Size and ζ-potential measurement of silica nanoparticles in serum using tunable resistive pulse sensing. *Langmuir* 32: 2216–2224. https://doi.org/10.1021/acs.Langmuir.5b04160.

59 Filipe, V., Hawe, A., and Jiskoot, W. (2010). Critical evaluation of nanoparticle tracking analysis (NTA) by insight to measure nanoparticles and protein aggregates. *Pharm. Res.* 27: 796–810. https://doi.org/10.1007/s11095-010-0073-2.

60 Gross, J., Sayle, S., Karow, A.R. et al. (2016). Nanoparticle tracking analysis of particle size and concentration detection in suspensions of polymer and protein samples: influence of experimental and data evaluation parameters. *Eur. J. Pharm. Biopharm.* 104: 30–41. https://doi.org/10.1016/j.ejpb.2016.04.013.

61 Cho, C.H., Aspetti, C.O., Park, J., and Agarwal, R. (2013). Silicon coupled with plasmon nanocavities generates bright visible hot luminescence. *Nat. Photonics* 7: 285–289.

62 Chowdhury, F.I., Nayfeh, M.H., and Nayfeh, A.M. (2016). Enhanced performance of thin-film silicon solar cells with a top film of silicon nanoparticles due to down-conversion and near resonance charge transport. *J. Sol. Energy* 125: 332–338.

63 Swinehart, D.F. (1962). The Beer-Lambert law. *J. Chem. Educ.* 39: 333. https://doi.org/10.1021/ed039p333.

64 Peng, K., Fu, L., Yang, H., and Ouyang, J. (2016). Perovskite $LaFeO_3$/−montmorillonite nanocomposites: synthesis, interface characteristics and enhanced photocatalytic activity. *Sci. Rep.* 6: 19723. https://doi.org/10.1038/srep19723.

65 Eustis, S. and El-Sayed, M.A. (2006). Why gold nanoparticles are more precious than pretty gold: noble metal surface plasmon resonance and enhancement of the radiative and nonradiative properties of nanocrystals of different shapes. *Chem. Soc. Rev.* 35: 209–217. https://doi.org/10.1039/B514191E.

66 Khlebtsov, N. and Dykman, L. (2010). Optical properties andbiomedical applications of plasmonic nanoparticles. *J. Quant. Spectrosc. Radiat. Transf.* 111: 1–35. https://doi.org/10.1016/j.jqsrt.2009.07.012.

67 Khlebtsov, N.G. and Dykman, L.A. (2010). Optical properties and biomedical applications of plasmonic nanoparticles. *J. Quant. Spectrosc. Radiat. Transfer* 111: 1–35. https://doi.org/10.1016/j.jqsrt.2009.07.012.

68 Reiss, G. and Hutten, A. (2005). Magnetic nanoparticles: applications beyond data storage. *Nat. Mater.* 4: 725–726. https://doi.org/10.1038/nmat1494.

69 Faivre, D. and Bennet, M. (2016). Materials science: magnetic nanoparticles line up. *Nature* 535: 235–236. https://doi.org/10.1038/535235a.

70 Qi, M., Zhang, K., Li, S. et al. (2016). Superparamagnetic Fe_3O_4 nanoparticles: synthesis by a solvothermal process and functionalization for a magnetic targeted curcumin delivery system. *New J. Chem.* 4480: 4480–4491. https://doi.org/10.1039/c5nj02441b.

71 Wu, W., He, Q., and Jiang, C. (2008). Magnetic iron oxide nanoparticles: synthesis and surface functionalization strategies. *Nanoscale Res. Lett.* 3: 397–415. https://doi.org/10.1007/s11671-008-9174-9.

72 Guo, D., Xie, G., and Luo, J. (2014). Mechanical properties of nanoparticles: basics and applications. *J. Phys. D: Appl. Phys.* 47: 13001. https://doi.org/10.1088/0022-3727/47/1/013001.

73 Lee, S., Choi, S.U.-S., Li, S., and Eastman, J.A. (1999). Measuring thermal conductivity of fluids containing oxide nanoparticles. *J. Heat Transfer* 121: 280–285. https://doi.org/10.1115/1.2825978.

74 Cao, Y.C. (2002). Nanoparticles with Raman spectroscopic fingerprints for DNA and RNA detection. *Science* 80 (297): 1536–1540. https://doi.org/10.1126/science.297.5586.1536.

75 Loureiro, A., Azoia, N.G., Gomes, A.C., and Cavaco-Paulo, A. (2016). Albumin-based nanodevices as drug carriers. *Curr. Pharm. Des.* 22: 1371–1390.

76 Alexis, F., Pridgen, E., Molnar, L.K., and Farokhzad, O.C. (2008). Factors affecting the clearance and biodistribution of polymeric nanoparticles. *Mol. Pharmaceutics* 5: 505–515. https://doi.org/10.1021/mp800051m.

77 Ali, A., Zafar, H., Zia, M. et al. (2016). Synthesis, characterization, applications, and challenges of iron oxide nanoparticles. *Nanotechnol. Sci. Appl* 9: 49–67. https://doi.org/10.2147/NSA.S99986.

78 Jain, P.K., Lee, K.S., El-Sayed, I.H., and El-Sayed, M.A. (2006). Calculated absorption and scattering properties of gold nanoparticles different size, shape, and composition: applications in biological imaging and biomedicine. *J. Phys. Chem. B* 110: 7238–7248. https://doi.org/10.1021/jp057170o.

79 Calvo, P., Remuoon-Lopez, C., Vila-Jato, J.L., and Alonso, M.J. (1997). Novel hydrophilic chitosan-polyethylene oxide nanoparticles as protein carriers. *J. Appl. Polym. Sci.* 63: 125–132. https://doi.org/10.1002/(SICI)1097-4628(19970103)63:1*125::AID-APP13*3.0.CO;2-4.

80 Laurent, S., Forge, D., Port, M. et al. (2010). Magnetic iron oxide nanoparticles: synthesis, stabilization, vectorization, physicochemical characterizations, and biological applications. *Chem. Rev.* 110: 2064–2110. https://doi.org/10.1021/cr900197g.

81 Zhang, J. and Saltzman, M. (2013). Engineering biodegradable nanoparticles for drug and gene delivery. *Chem. Eng. Prog.* 109: 25–30.

82 Prashant, K.J. and Ivan, H.S. (2007). Au NPs target cancer. *Nano Today* 2: 19–29.

83 Chen, C., Xing, G., Wang, J. et al. (2005). Multihydroxylated $[Gd@C_{82}(OH)_{22}]_n$ nanoparticles: antineoplastic activity of high efficiency and low toxicity. *Nano Lett.* 5: 2050–2057. https://doi.org/10.1021/nl051624b.

84 AshaRani, P.V., Low Kah Mun, G., Hande, M.P., and Valiyaveettil, S. (2009). Cytotoxicity and genotoxicity of silver nanoparticles in human cells. *ACS Nano* 3: 279–290. https://doi.org/10.1021/nn800596w.

85 Hajipour, M.J., Fromm, K.M., Ashkarran, A.A. et al. (2012). Antibacterial properties of nanoparticles. *Trends Biotechnol.* 30: 499–511. https://doi.org/10.1016/j.tibtech.2012.06.004.

86 Yin, Q., Wu, W., Qiao, R. et al. (2016). Glucose assisted transformation of Ni-doped-ZnO@carbon to a Ni-dopedZnO@void@SiO_2 core–shell nanocomposite photocatalyst. *RSC Adv.* 6: 38653–38661. https://doi.org/10.1039/C5RA26631A.

87 Todescato, F., Fortunati, I., Minotto, A. et al. (2016). Engineering of semiconductor nanocrystals for light-emitting applications. *Materials* 9: 672. https://doi.org/10.3390/ma9080672.

88 Weiss, J., Takhistov, P., and McClements, D.J. (2006). Functional materials in food nanotechnology. *J. Food Sci.* 71: R107–R116. https://doi.org/10.1111/j.1750-3841.2006.00195.x.

89 Lei, Y.-M., Huang, W.-X., Zhao, M. et al. (2015). Electrochemiluminescence resonance energy transfer system: mechanism and application in ratiometric aptasensor for lead ion. *Anal. Chem.* 87: 7787–7794. https://doi.org/10.1021/acs.analchem.5b01445.

90 Unser, S., Bruzas, I., He, J., and Sagle, L. (2015). Localized surface plasmon resonance biosensing: current challenges and approaches. *Sensors* 15: 15684–15716. https://doi.org/10.3390/s150715684.

91 Ripp, S. and Henry, T.B. (eds.) (2011). *Biotechnology and Nanotechnology Risk Assessment: Minding and Managing the Potential Threats Around Us*, ACS Symposium Series. Washington, DC: American Chemical Society http://dx.doi.org/10.1021/bk-2011-1079.

92 Golobič, M., Jemec, A., Drobne, D. et al. (2012). Upon exposure to Cu nanoparticles, accumulation of copper in the isopod *Porcellio scaber* is due to the dissolved Cu ions inside the digestive tract. *Environ. Sci. Technol.* 46: 12112–12119. https://doi.org/10.1021/es3022182.

93 Swadeshmukul, S., Peng, Z., Kemin, W. et al. (2001). Conjugation of biomolecules with luminophore-doped silica nanoparticles for photostable biomarkers. *Anal. Chem.* 73: 4988–4993. https://doi.org/10.1021/AC010406+.

94 Tratnyek, P.G. and Johnson, R.L. (2006). Nanotechnologies for environmental cleanup. *Nano Today* 1: 44–48. https://doi.org/10.1016/S1748-0132(06)70048-2.

95 Mueller, N.C. and Nowack, B. (2008). Exposure modeling of engineered nanoparticles in the environment. *Environ. Sci. Technol.* 42: 4447–4453. https://doi.org/10.1021/es7029637.

96 Rogozea, E.A., Petcu, A.R., Olteanu, N.L. et al. (2017). Tandem adsorption-photodegradation activity induced by light on NiO-ZnO *p–n* couple modified silica nanomaterials. *Mater. Sci. Semicond. Process.* 57: 1–11. https://doi.org/10.1016/j.mssp.2016.10.006.

97 Rogozea, E.A., Olteanu, N.L., Petcu, A.R. et al. (2016). Extension of optical properties of ZnO/SiO_2 materials induced by incorporation of Au or NiO nanoparticles. *Opt. Mater.* 56: 45–48. https://doi.org/10.1016/j.optmat.2015.12.020.

98 Kosmala, A., Wright, R., Zhang, Q., and Kirby, P. (2011). Synthesis of silver nanoparticles and fabrication of aqueous Ag inks for inkjet printing. *Mater. Chem. Phys.* 129: 1075–1080. https://doi.org/10.1016/j.matchemphys.2011.05.064.

99 Cushing, B.L., Kolesnichenko, V.L., and O'Connor, C.J. (2004). Recent advances in the liquid-phase syntheses of inorganic nanoparticles. *Chem. Rev.* 104: 3893–3946. https://doi.org/10.1021/cr030027b.

100 O'Brien, S., Brus, L., and Murray, C.B. (2001). Synthesis of monodisperse nanoparticles of barium titanate: toward a generalized oxide nanoparticle synthesis strategy. *J. Am. Chem. Soc.* 123: 12085–12086. https://doi.org/10.1021/ja011414a.

101 Wang, D.W. and Su, D. (2014). Heterogeneous nanocarbon materials for the oxygen reduction reaction. *Energy Environ. Sci.* 7: 576. https://doi.org/10.1039/c3ee43463j.

102 Wang, Z., Pan, X., He, Y. et al. (2015). Piezoelectric nanowires in energy harvesting applications. *Adv. Mater. Sci. Eng.* 2015: 1–21. https://doi.org/10.1155/2015/165631.

103 Kot, M., Major, Ł., Lackner, J.M. et al. (2016). Mechanical and tribological properties of carbon-based graded coatings. *J. Nanomater.* 2016: 1–14. https://doi.org/10.1155/2016/8306345.

2

Implications of Nanotechnology and Environment

CHAPTER MENU

2.1 Ecotoxicological Implications of Nanoparticles

Substantial quantities of engineered nanoparticles have been introduced into the atmosphere by creating nanotechnologies, including those in ambient air and water. An increasing number of studies have concentrated on assessing the toxicity of nanoparticles commonly used in the industry to protect human health and biodiversity from the potentially harmful effects of a wide variety of nanoparticles [1].

Oxidative stress is convenient to quantify in the sense of toxicity and ecotoxicity because cells react with many defensive responses, easily counted as an enzyme or genetic reaction, to oxidative stresses [2]. Several in vitro toxicity tests of nanoparticles, for example, TiO_2 and fullerenes, have shown that reactive oxygen species (ROS) is generated [3, 4]. This deceptive contrast illustrates the necessity in animals and individual cells to research nanoparticles' interactions and nanoparticles' mechanistic aspects.

Many nanoparticles are photochemically active to produce excited electrons to the sun. When oxygen is present, super radicals can be formed by direct transmission of electrons [5], in an ecotoxicity sense, circumstances in which species are simultaneously exposed to nanoparticles and light.

Ecotoxicity assessments are carried out at a range of trophic stages, including certain species and some exposure conditions, such as microorganisms, plants, invertebrates, and vertebrates. The structured protocols describe individual physical and physiological states and direction and growth and fertility. The benefits of microorganisms are all-round and very diverse, tiny, and short generation [6].

Nano-Technological Intervention in Agricultural Productivity, First Edition.
Javid A. Parray, Mohammad Yaseen Mir and Nowsheen Shameem.

Apart from microorganisms, there are many very useful organisms from different environments that may be used in ecotoxicity testing. Organisms of different trophic levels from primary producers to predators can be of concern because they accumulate in the food chain (bioamplification). This is because of environmental contaminants. Organisms can also be chosen depending on multiple habitats in the soil and water because they become essential to the biogeochemical cycling of elements [5] for carrying out an environmental method. Ecotoxicity means toxicity exclusively for species of ecological concern. Simultaneously, the term bioassay requires the assessment of toxicity or stress by a substance within an ecological matrix that applies to live in the nature of the organism.

The exposure of fish to nanoparticles in pure water can, therefore, constitute an ecotoxicity trial while exposure of fish to nanoparticles in saltwater results in dissolution of organic carbon (DOC) [6]. Changes in nanoparticles such as ions, natural colloids, and other locked surfaces as they come in contact with the environmental matrix components would likely affect not only mobility, aggregation, and so on. Still, they will also change the toxicity properties [7]. Ecotoxicology, as in traditional toxicology, can also use a wide range of physiological and genetic endpoints. The functional testing of ecotoxic species nevertheless contributes to the difficulty of such studies. Very pure systems have been used in most research on nanoparticulate adverse effects, where nanoparticles can interact with matrix constituents that affect bioavailability [6].

The safety doubts of NPs were emphasized, and their use was investigated both purposely and accidentally concerning the potential dangers associated with NPs [8, 9]. Massive industrial processing of nanoparticles might soon lead to the presence, in different environments and humans, of both nanoparticles and waste produced when presenting them. Once you evaluate nanoparticles' hazards from the environmental perspective, there is a paradox that potentially harmful nanoparticles often possess the potential of creating the so-called green chemistry. These more environmentally friendly processes can be used in ecological contaminants [10]. Examples are the use of engineered nanoparticles to treat water and fix the water, which has proven successful and has also created questions about human exposures to nanoparticles in treated water. Research to determine whether or not a substance is harmful includes assessing the material's inherent toxicity and its interaction with living cells [11]. In in vitro and in vivo toxicological tests, the doses or dosage levels used are most often too high relative to actual accidental human exposure [8].

More research is therefore needed before widespread statements on ecotoxicological nanoparticles could be produced. The best approach subject to scrutiny is to prevent the escape of nanoparticles to the atmosphere. This decreases their mobility and prevents their presence in the atmosphere by inserted NPs into organic or inorganic matrices [12]. The easiest way to improve the protection of NPs is the use of nanocomposites like these. The use of magnetic nanoparticles in its architecture is a supportive approach to ensuring nanoparticles' security [13]. Based on the useful characteristics and the response to a magnetic field, vegetable NPs are prodigious influential scientists [14]. Polymeric material containing magnetic nanoparticles with specific characteristics, catalytic or bactericidal operation fields, is that magnetic nanoparticles can be retrieved quickly using simple magnetic traps to leak the nanocomposite [15]. Nanoparticles' ecotoxic consequences are as follows.

2.1.1 Ecotoxicity of Fullerenes

C60 fullerenes, beginning with the seminal work of Oberdorster [16], have been among the first engineered nanoparticles to be studied in ecotoxicity. In the first study on fullerene toxicity [16], Oberdorster showed that low C60 (0.5 mg/l) concentrations could result in oxidation and enzyme changes. The C60 model influences the ecotoxicity of C60. Bacterial studies of C60 fullerenes showed reduced growth and breathing with *Escherichia coli* and *Bacillus subtilis* [17, 18]. Other reports demonstrated a responsive cellular property impaired by low nC60 concentration [14] in bacterial membrane composition and fluidity. In the presence of nC60, *Pseudomonas putida* reduced its unsaturated fatty acids and augmented cyclopropane fatty acids to defend the bacterial membrane against oxidative stress.

Changes in the composition of the population revealed growth barriers to some indigenous microbial organisms. They showed that the ecotoxicological effects of nC60 could occur within the highly complex medium, including soils. Without previous suspension in water, the maximum concentration of C60 was added in dry conditions. The lack of any ecotoxicological effects may suggest that considering higher concentrations, the bioavailability of C60 in the latter experiment is lower [6].

Fullers (1 μM) combined with UV light have shown a substantially larger inactivation (100% increase) of viruses than UV light alone. In a cytotoxicity analysis on human cell lines [3], fullerols are less toxic than aggregate non-derivatized fullerenes (nC60). C60-NH_2 with a positive charge hindered the growth and reduced the absorption of substrates. This damage has been recorded in two bacterial species (*E. coli* and *Shewanella oneidensis*), at a concentration of 10 mg/l, on the cell structures (cell walls and membranes) [19]. The same study showed moderate to no adverse effects for neutrally charged C60 and C60-OH [20].

2.1.2 Ecotoxicity of Carbon Nanotubes

The ecotoxicity of zebras (*Danio rerio*) under various salinity conditions for unprocessed, one-, and double-wall carbon nanotubes. In zebrafish embryos, single-walled carbon nanotubes (SWCNTs) have substantial delays in hatching at concentrations above 120 mg/l. Two-walled carbon nanotubes also provided a hatching delay of more than 240 mg/l. Carbon black did not affect hatching, however. SWCNT concentrations of up to 360 mg/l [21] have not affected embryo development. In the Rainbow Trout experiment, prepared with surfactant sodium dodecyl sulfate (SDS) and sonicity [22], a dose-dependent increase in ventilation rates and gill pathology (oedema, altered monocytes, and hyperplasia) at concentrations between 0.1 and 0.5 mg/l was observed. The authors concluded that the trout's SWCNTs might act as a respiratory toxicant and precipitate gill mucus. However, this did not have a significant effect on death, development, or reproduction [23].

2.1.3 Ecotoxicity of Metal Nanoparticles

Metal toxicity is affected by different factors, such as solubility, biological site specificity, and the like. Toxic effects and heavy metal toxicity are defined as any functional or morphological body changes resulting from ingested, injected, inhaled, or consumed medications and chemical or biological agents [24, 25]. There are also antibacterial processes in metal nanoparticles. Depending on the load on the membrane surface, metal nanoparticles practise cytotoxicity. Gram-positive cells are less vulnerable to genotoxic effects because of a thicker peptidoglycan layer compared to Gram-negative cells. Nanotoxicity can be due to the energy exchange between membrane nanoparticles and their aggregation in the cytoplasm [26]. In dechlorinated water with a hardness of 142 mg $CaCO_3$/l and a pH of 8.2 [27], copper nanoparticles were tested for comparison with the toxicity of soluble Cu ions in zebrafish (*D. rerio*) [27] ($CuSO_4$). In this sample, Cu nanoparticles were lower in toxicity than Cu ions. Like Ag [28], nanoparticles are used as an antimicrobial agent. Silver nanoparticles play an essential role in the field of nanotechnology and nanomedicine. Their specific size-dependent characteristics make these peculiar physical, chemical, and biological properties superior and indispensable. The possible antimicrobial activity of silver nanoparticles is numerous pathogenic microorganisms.

Silver nanoparticles demonstrate undesirable toxic effects on human health and the environment alongside this antimicrobial activity [1]. Ecologists have cautioned that nano-antimicrobials released into water sources. The widespread release of such a potent antimicrobial may have significant negative implications for bacteria in natural systems. There is also increasing evidence that silver nanoparticles and being toxic to bacteria are incredibly harmful to mammalian cells [29]. Silver nanoparticles damage brain cells [30], liver cells [31], and stem cells. Silver is also extremely toxic to fish [32] algae, individual plants, fungi [33] crustaceans, and bacteria such as heterotrophic nitrogen-fixing bacteria and chemical soil-forming bacteria [34] in its bulk form. Silver nanoparticles, which have well-known bactericidal and cytotoxic effects, including specific mitochondrial products and ROS generation, are mainly produced for antiseptic purposes [31].

Silver is also extremely toxic to fish [32] algae, individual plants, fungi [33] crustaceans, and bacteria such as heterotrophic nitrogen-fixing bacteria and chemical soil-forming bacteria [34] in its bulk form. Silver nanoparticles, including unique mitochondrial products and ROS generation, are mainly formulated for antiseptic purposes and have well-established bactericidal and cytotoxic effects [31, 35].

2.1.4 Ecotoxicity of Nanocomposites

Nanocomposites vowed to do so because of their multifunction, which means that distinctive properties with conventional materials can be combined [36]. Depending on the nanophase and matrix nature, a wide variety of nanocomposites can be developed [37]. Through two separate viewpoints, the concept of using polymer metal nanocomposites is beneficial. First, by preventing the autocomposition of polymer-stabilized metal nanoparticles, nanoparticles' production can be considered one of the most encouraging resolutions to stability. Furthermore, the use of immobilized NPs limits their environmental changes [38, 39]. The features of the metal nanoparticles do not primarily determine the characteristics of nanocomposites. For example, the formulation of metal nanoparticles in polymer matrices can significantly change the polymer morphology because of the existence of nanoporosity, which promotes mass transfer within the nanocomposites and specific additional structural parameters that are of high significance in their practical applications [40]. A significant number of water purification and efficiency problems are solved by nanotechnology [41]. The application of metal nanoparticles in the field of reductive dechlorination of organic halogenated compounds in groundwater has been meticulously observed [42]. As pure monometallic entities or in combination with platinum, iron nanoparticles are one of the most substantial components. The long-term stability of these nanoparticles can, however, be enhanced by immobilization in stable support. The ion sharing is widely used in various water treatment ion-sharing processes to prevent undesired or harmful ionic impurities such as hardness ions, iron, and heavy metals. The modification of such bactericidal metal nanoparticles allows for eliminating microbiological contaminants by a combination of traditional water treatment with disinfection. One person may perform the two extra water treatment measures with one person.

Nanoparticles of titanium dioxide (TiO_2) tend to be of little toxicity to terrestrial species and are used as nanocomposites on a sunscreen where TiO_2 has been coated with magnesium, silica, or alumina and also with amphiphilic organic substances such as polydimethylsiloxane (PDMS), which ageing alters [43].

Another critical technical problem, known as biofouling, can also be solved using silver nanoparticles containing nanocomposites. Biofouling or biological fouling [37] is the unintended accumulation of microbes on the surface of water treatment systems and materials, such as reverse osmosis membranes, cooling water cycles, and ion exchange resins. The surface shift technique was based on commercially available ion exchange materials with a silver shell [42] and a magnetic core [43]. These products are the eco-friendly bactericide nanocomposites suited for traditional water treatment and reagent-free disinfection. These materials have the main advantages as follows: (i) They are trapped firmly into the polymer matrix, which prohibits the escape into the medium being processed. (ii) Applying a material's surface metallic nanoparticles ensures interaction with the bacteria to avoid fast water disinfection. (iii) Metal nanoparticles are superparamagnetically shielded in nature because a simple magnet trap prevents any post-contamination of treated water with metal nanoparticles leached by the polymer matrix. (iv) Like the ion exchange capacity [44], the surface location of metal nanoparticles does not essentially affect the core characteristics of ion exchange materials.

2.1.5 Ecotoxicity of Oxide Nanoparticles

TiO_2, SiO_2, and ZnO nanoparticles have been examined in water suspensions of citrate containing low concentrations of PO_4 in Gram-positive and harmful bacteria such as *B. subtilis* and *E. coli* [44]. These effects are generally most susceptible to *B. subtilis* by operations produced from SiO_2 to TiO_2 to ZnO [6]. Magnetic nanoparticles are of great interest to researchers in many areas, including catalysis, bioengineering/biomedicine, and environmental science and technology [45]. Concerning material protection, it is essential to remember that ferrites are very suitable for biological purposes because of their low toxicity and strong magnetic characteristics [46]. In catalysis or water treatment, core/shaft metal nanoparticles or nanocomposites can be added [47]. For instance, in catalysis, nanocomposites can be quickly recovered and reused by a sequence of catalytic cycles because of the magnetic properties of NPs comprising platinum metals [40].

Zn and ZnO are phytotoxic to germination of seeds and root development following two hours in exposure to deionized water nanoparticle suspensions [48]. There were screening of five types of nanoparticles (corn and cucumber) and six plant species (radish, rape, and ryegrass). Approximately 50% inhibition of root growth was observed in nano-Zn and nano-ZnO at about 50 mg/l for radish and about 20 mg/l for rape and ryegrass. Reports indicate that pure alumina particles significantly reduce radical elongation in all plant species, potentially slowing plant growth. Alumina nanoparticles may be emitted into the atmosphere by exhaust systems and combined with other airborne materials. Alumina was also packed with phenanthrene, an essential element of polycyclic aromatic hydrocarbons in the atmosphere that can be absorbed into a particular substance in the air. They substantially reduced their phytotoxicity without having any harmful effects on the roots of plants [1].

2.2 Nanotechnology and Agriculture

There is great interest in nanotechnology's potential advantages in farming to improve agricultural production with little energy and expenditure inputs. The significant possible applications under nanoparticles in the farm sector were nanosensor systems, nanopesticides and nanofertilizers, and smart agrochemicals supply systems [49]. Besides, plant breeding and genetic engineering devices based on the nanotechniques [50] are also used. The development of nanotechnological approaches for farming, especially crop productivity and disease management, was encouraged [51]. Pesticide encapsulation of nanoparticles for continuous release, nanoparticles mediated by genetics for crop improvement, carbon nanoparticles supported by rain-fed seed germination plants, nanofertilizers for improved crop nutrition and crop productivity, and nanopesticide management of plant diseases; De La Torre-Roche et al. [52] indicates a decrease in the use of plant pesticides in combination with nanomaterials. This result demonstrates how these interactions benefit from reducing the residues of pesticides in plants and other parts.

In an age when increased supply for sustainability is bound to minimize the price and overuse of agricultural and natural resources, nanotechnology offers the possibility of precision cultivation [36]. However, considering the promising results achieved through the involvement in agriculture of advanced nanotechnologies, their importance to the market has not yet been reached [53]. Furthermore, fluctuating suspicions regarding the outcome, transport, bioavailability, and toxicity of NPs restrict the full acceptance and readiness of agricultural nanotechniques.

Consequently, nanotechnology provides a precise capacity to revitalize the agricultural industry. It is worth noting at the same time; nevertheless, that specific contribution to agriculture still is unclear and increasing. Therefore, a system-level approach is essential to provide more precise information about the exposure of nanoparticles and their risk in agricultural systems. In the future, deciding factors are future guidance for improved forestry nanoparticle science, which will focus on optimizing the safe use of nanoparticles at an appropriate level of farming gain in modulating fate, actions, bioavailability, and toxicity.

2.3 Risk Assessment Factors and Modulation of Nanomaterials

The present situation shows the successful progress and renovating of the agricultural sector by the exciting science of nanotechnology applications. Such improvements are ultimately desperately needed to feed the rising global population, likely to reach nine billion by 2050. Because of the potential benefits of agricultural nanotech products soon, many countries worldwide are making a significant effort to assess the adequacy and feasibility of nanotechnology incorporation with cultivation. After nanoparticles are discharged into the atmosphere, they can interact with these parts of the agroecosystem [53]. Therefore, researchers aim to understand and track these essential interactions' essence to obtain realistic knowledge about the toxicity and potential impacts of released nanoparticles on the environment and agriculture. These investigations would also contribute mainly to determining the appropriate quantity of nanoparticles at all safe amounts tolerated.

Any changes in the plant–soil relationship, the main driver of agricultural production, influence the soil system's physicochemical properties. In particular, the soil is the primary sink of processed nanoparticles. Their next contact with various soil components may thus get an important effect on nanoparticles' fate, transportation, and behaviour. For instance, the significance of soil pH, organic matter material, and capacity to monitor their fate, toxicity, and bioavailability have been pointed out in few reports of most common, most studied, silver nanoparticles (AgNPs), with antimicrobial properties [54, 55] such as lesser soil pH, organic substance content, and the process of exchanging cation disrupt Ag to soil solubility. It leads to increased risk of AgNPs being mobile, toxic, or bioavailable. Instead, higher soil pH, organic matter content, and cation exchange capacity allow Ag to be soil sorbed, preventing AgNP mobility, bioavailability, and additional toxicity [55]. Wang et al. [56] have conducted a comparable toxicity study of ZnO-NPs in soil and solution farming. The authors called for significantly decreasing the toxicity of ZnO-NPs in the soil environment. It is mainly due to various soil properties that assess Zn's phytotoxicity of the soil systems, while toxicity is the dissolution of particles in solution cultures. Also, a 2000 mg/kg loamy sand concentration of pH 5.5 was reported not to show any phytotoxicity of ZnO-NPs to *Cucumis sativus* [57, 58].

Besides, soil organic matter is seen as another significant factor affecting ZnO-NP transportation behaviour, which eventually determines its toxicity. In reducing the toxicity of ZnO-NPs in the direction of *Zea mays*, Zhao et al. [59] demonstrated the positive influence of alginate. Besides, ZnO-NPs showed no reduction in plant biomass in the earth with alginate at 400–800 mg/kg while a substantial decrease was noted in biomass production without alginate.

The nanoparticles injected into the atmosphere gradually accrue into the soil, with soil types mainly influenced by their origin, transport, and behaviour [60]. Ge et al. [61] have observed, in a dose-dependent way, the possible impact of TiO_2 and ZnO-NPs on the soil bacterial population. By DNA-based fingerprinting study, reduced bacterial diversity has been reported, responding to these nanoparticles, decreasing Rhizobial, Bradyrhizobial, and Bradyrhizobiaceae taxa.

TiO_2 and ZnO-NPs have modified the bacteria structure substantially, with a distinct environmental effect. For example, the falling taxa are closely related to the process of N_2 fixation, while growing taxa are likely to affect organic and biopolymer decomposition processes. Furthermore, Shahrokh et al. [62] have found that the dose-dependent effects of AgNPs on Rhizobium and Azotobacter nitrate reductase activities were facilitated by the low doses of AgNPs in Azotobacter (0.2 ppm).

Depending on these experiments' outcomes, the denitrifying bacterial population has been predicted to be highly sensitive to toxicity to nanoparticles [63]. Despite the well-known effect of nanoparticles on soils' microbial community, there is a lack of literature that links soil element explicitly to toxic comportment of nanoparticles to the biota of soil [54]. Frenk et al. [64] also showed the effects on soil bacteria of two different types of soil of copper oxide (CuO) and magnetite (Fe_3O_4) nanoparticles.

It is important to note that there were more adverse effects from both nanoparticles in sandy loam soil, with CuO having a relatively strong impact on the composition and bacteria of the culture. CuO, affected more precisely, is the most focused taxa in the sandy loam soil of Rhizobial and Sphingobacteriaceae. Based on this observation, the incidence of a clay segment in the soil has decreased the nanoparticle toxicity. The ecotoxicology of ZnO-NPs on soil microbes was illustrated [65] based on parameters such as ammonification, breathing, dehydrogenase, and fluorescent diacetate hydrolase activity. In acidic and neutral soils, the adverse effects of ZnO-NPs on microcosmic soil microbes have been found to have a more significant impact.

Alkaline soil, on the other hand, was very damaging. The toxicity of TiO_2 NPs was also determined mainly to be influenced by soil and organic matter pH [66]. This substantially reduced carbon in high pH and organic soils. These findings show that nanoparticles show acute toxicity to soil microbial substances in the soil environment to define important soil parameters such as form, organic soil matter, and soil humidity. Significantly, intimate description of the possible effect of different soil properties on the toxicity of the nanomaterials helps to reduce the fuel toxicity of nanoparticles produced throughout the soil environment.

Nanoparticles' plant phytotoxicity is yet another concern because of the leaves, and root system plants' complete surface area has enough opportunity to interact with nanoparticles. Moreover, because their minuscule size can translocate effectively inside the host plant, the risk of nanoparticles becoming toxic is higher in plants. Nanoparticles in the plant body are assumed to enter by surface adsorption or pass through small plant openings [67, 68]. NPs depend on their size and concentration primarily on plant toxicity. Furthermore, thanks to plants' easy absorption and their subsequent translocation within the system, small nanoparticles can cause phytotoxic toxicity at even lower levels [56]. Most commonly used metal nanoparticles, i.e. AgNPs, are thought to play a crucial role in their phytotoxic behaviour, the scale of nanoparticles [69].

Asli and Neumann [70] also noted TiO_2 NPs' mechanical treatment concentration mode for hydraulic leakage and Z transpiratory rate. The hydroponic solution is cultivated in mays. In contrast, Yang et al. [71] noticed the stimulating impact of TiO_2-NPs (2.5 g/l) application on fresh and dry weight of *Spinacia oleracea* and also elevated response was seen in case of chlorophyll, protein and total nitrogen content in leaves. Also in cultivations of cultivated media with a low concentration of 0.5 g/l supplements of TiO_2-NPs, Song et al. [72]

Table 2.1 Nanomaterials and their impacts on plants and soil microbiota.

Soil characteristics	Nanoparticles	Impacts	References
Soil type			
Silt and clay soil	Titanium oxide	Substantially limited mineralization of carbon	[66]
Loam sandy	Titanium oxide	Affects soil microbiota	[66]
	Zinc oxide	No *Cucumis sativus* toxicity with soil pH 5.5 at a concentration of 2000 mg/kg	[57]
	Cupric oxide, zinc oxide	The toxic effect on *Triticum aestivum*	[82]
	Silver	Reduced microbial biomass	[83]
	Silver	Significantly lower soil enzymatic activities and respiration caused by the substrate	[84]
	Cerium oxide, tin oxide, zinc oxide	Microbial biomass C and N no effect	[85]
	Titanium oxide	Reduced bacterial diversity	[86]
	Titanium oxide	Reduced bacterial taxa	[61]
	Amine-modified polystyrene nanospheres and TiO_2 nanoparticles	Decreases in rhizosphere bacterial counts and plant root and stem growth	[87]
	Sulfate-modified polystyrene nanospheres	Increased rhizosphere bacterial counts	[87]
Clay and loamy	Zinc oxide	Toxic impact on *Triticum aestivum*	[58]
pH			
Acidic	Silver, zinc oxide	Increased toxicity to *Eisenia fetida*	[65]
Alkaline	Titanium oxide	The decrease in the microbial population in soil	[66]
	Silver	Declined toxicity	[88]
Organic part			
High	Silver	Toxicity reduction for biofilm-forming populations	[89]
	Titanium oxide	Effect on C mineralization	[66]
	Zinc oxide	Significant effect of adding alginate at a concentration of 400–800 mg/kg/kg on *Zea mays*	[59]
Low	Cupric oxide, zinc oxide	Improved toxicity to microbes	[64]
Cation exchange capacity			
High	Silver	Decreased toxic impact on *Pseudomonas chlororaphis*	[90]
	Zinc oxide	No effect on *Lepidium sativum*	[81]
Low	Silver	Improved toxicity to soil microbes	[90]

have documented a positive effect on duckweed (*Lemna minor*); the also increased a concentration caused significant crop harm.

ZnO-NPs are by far the most widely used material oxide nitrogen NPs, and elevated *Vigna radiata* and *Cicer arietinum* have been documented to increase growth, comparatively, in the vegetal agar media [73]. The similar dose-based effect responses had been observed for Cu-NPs, where wheat (*Triticum aestivum*) and mung bean (*Phaseolus radiatus*) have significantly inhibited growth [74–78]. The critical route that plants are predominantly exposed to the released nanoparticles is the soil. It should be noted that many investigators have also begun to take this approach into account by avoiding the hydroponic system, which offers more appropriate toxicity data and straightforward understanding. Zhu et al. [79] also interpreted the soil system's nanofilling potential by preventing Cucurbita maximum uptake of Fe_3O_4-NPs. Fe_3O_4-NPs (20 nm) were traced into the leaves simultaneously when the plants were grown in a hydroponic growth process. Similarly, when subjected to soil irrigation, nanoparticles of CeO_2 had a neutral effect on maize plants [79].

On the other hand, the toxicity of AgNPs in *P. radiatus* and *Sorghum bicolor* cultivated in the soil system was lacking compared to agar [80]. Most notably, soil parameters are primarily used to regulate plant reactions to soil-released nanoparticles [81]; for instance, they have demonstrated the non-toxic effect of ZnONPs in the cultivation of *Lepidium sativum* with a higher cation exchange level. Also, CuO and ZnO nanoparticles were more toxic for *T. aestivum* [82]. Such results indicate which soil parameters can attenuate nanoparticles' likely phytotoxicity. The analysis of key risk assessment factors in the tripartite interaction of nanoparticles with microbial communities of plants, soils, and soils would understand the environmental effects of nanoparticles released into the agroecosystem (Table 2.1). Indeed, essential soil factors are mainly controlled by soil type, pH, organic matter, soil moisture, behaviour, and toxicity of nanoparticles manufactured for plants and microbes.

References

1 Yang, L. and Watts, D.J. (2005). Particle surface characteristics may play an important role in phytotoxicity of alumina nanoparticles. *Toxicol. Lett.* 158 (2): 122–132.

2 Kovochich, M., Xia, T., Xu, J. et al. (2005). Principles and procedures to assess nanoparticles. *Environ. Sci. Technol.* 39 (5): 1250–1256.

3 Sayes, C.M., Fortner, J.D., Guo, W. et al. (2004). The differential cytotoxicity of water-soluble fullerenes. *Nano Lett.* 4 (10): 1881–1887.

4 Daroczi, B., Kari, G., McAleer, M.F. et al. (2006). In vivo radioprotection by the fullerene nanoparticle DF-1 as assessed in a zebrafish model. *Clin. Cancer Res.* 12 (23): 7086–7091.

5 Hoffmann, M., Holtze, E.M., and Wiesner, M.R. (2007). Reactive oxygen species generation on nanoparticulate material. In: *Environmental Nanotechnology: Applications and Impacts of Nanomaterials* (eds. M.R. Wiesner and J.Y. Bottero), 155–203. New York: McGraw Hill.

6 Joner, E.J., Hartnik, T., and Amundsen, C.E. (2008). Environmental fate and ecotoxicity of engineered nanoparticles. In: *Norwegian Pollution Control Authority Report No. TA 2304/2007* (eds. E.J. Joner, T. Hartnik and C.E. Amundsen), 1–64. Norway: Bioforsk.

7 Lyon, D.Y., Thill, A., Rose, J., and Alvarez, P.J.J. (2007). Ecotoxicological impacts of nanomaterials. In: *Environmental Nanotechnology: Applications and Impacts of Nanomaterials* (eds. M.R. Wiesner and J.Y. Bottero), 445–479. New York: McGraw Hill.

8 Abbott, L.C. and Maynard, A.D. (2010). Exposure assessment approaches for engineered nanomaterials. *Risk Anal.* 30 (11): 1634–1644.

9 Maynard, A.D. (2007). Nanotechnology: the next big thing, or much ado about nothing? *Ann. Occup. Hyg.* 51 (1): 1–12.

10 Schulenburg, M. (2008). *Nanoparticles – Small Things, Big Effects Opportunities and Risks.* Berlin: Federal Ministry of Education and Research.

11 Bottero, J.Y., Rose, J., and Wiesner, M.R. (2006). Nanotechnologies: tools for sustainability in a new wave of water treatment processes. *Integr. Environ. Assess. Manage.* 2 (4): 391–395.

12 Macanas, J., Ruiz, P., Alonso, A. et al. (2011). Ion-exchange assisted synthesis of polymer-stabilized metal nanoparticles. In: *Solvent Extraction and Ion Exchange: A Series of Advances*, vol. 20 (ed. S.G. AK), 1–43. Boca Raton, FL: CRC Press-Taylor & Francis Group.

13 Vatta, L.L., Sanderson, R.D., and Koch, K.R. (2006). Magnetic nanoparticles: properties and potential applications. *Pure Appl. Chem.* 78 (9): 1793–1801.

14 Belotelov, V.I., Perlo, P., and Zvezdin, A.K. (2005). Magneto optics of granular materials and new optical methods of magnetic nanoparticles and nanostructures imaging. In: *Metal-Polymer Nanocomposites*, vol. 8 (eds. L. Nicolais and G. Carotenuto), 201–240. New York: Wiley.

15 Qiao, R., Zhang, X.L., Qiu, R. et al. (2007). Fabrication of superparamagnetic cobalt nanoparticles-embedded block copolymer microcapsules. *J. Phys. Chem. C* 111 (6): 2426–2429.

16 Oberdorster, E. (2004). Manufactured nanomaterials (fullerenes, C_{60}) induce oxidative stress in the brain of juvenile largemouth bass. *Environ. Health Perspect.* 112 (10): 1058–1062.

17 Fortner, J.D., Lyon, D.Y., Sayes, C.M. et al. (2005). C_{60} in water: nanocrystal formation and microbial response. *Environ. Sci. Technol.* 39 (11): 4307–4316.

18 Wiesner, M.R., Lowry, G.V., Alvarez, P. et al. (2006). Assessing the risks of manufactured nanomaterials. *Environ. Sci. Technol.* 40 (14): 4336–4345.

19 Tang, Y.J., Ashcroft, J.M., Chen, D. et al. (2007). Charge-associated effects of fullerene derivatives on microbial structural integrity and central metabolism. *Nano Lett.* 7 (3): 754–760.

20 Zhu, X., Zhu, L., Li, Y. et al. (2007). Developmental toxicity in zebrafish (*Danio rerio*) embryos after exposure to manufactured nanomaterials: buckminsterfullerene aggregates (nC_{60}) and fullerol. *Environ. Toxicol. Chem.* 26 (5): 976–979.

21 Cheng, J., Flahaut, E., and Shuk, H.C. (2007). Effect of carbon nanotubes on developing zebrafish (*Danio rerio*) embryos. *Environ. Toxicol. Chem.* 26 (4): 708–716.

22 Smith, C.J., Shaw, B.J., and Handy, R.D. (2007). Toxicity of single walled carbon nanotubes to rainbow trout, (*Oncorhynchus mykiss*): respiratory toxicity, organ pathologies, and other physiological effects. *Aquat. Toxicol.* 82 (2): 94–109.

23 Roberts, A.P., Mount, A.S., and Seda, B. (2007). In vivo biomodification of lipid-coated carbon nanotubes by *Daphnia magna*. *Environ. Sci. Technol.* 41 (8): 3028–3029.

24 Oberdorster, E., Zhu, S., Blickley, T.M. et al. (2006). Ecotoxicology of carbon-based engineered nanoparticles: effects of fullerene (C_{60}) on aquatic organisms. *Carbon* 44 (6): 1112–1120.

25 Panyala, N.R., Pena-Mendez, E.M., and Havel, J. (2008). Silver or silver nanoparticles: a hazardous threat to the environment and human health. *J. Appl. Biomed.* 6 (3): 117–129.

26 Rana, S. and Kalaichelvan, P.T. (2011). Antibacterial effects of metal nanoparticles. *Adv. Biotech* 2 (2): 21–23.

27 Griffitt, R.J., Weil, R., Hyndman, K.A. et al. (2007). Exposure to copper nanoparticles causes gill injury and acute lethality in zebrafish (*Danio rerio*). *Environ. Sci. Technol.* 41 (23): 8178–8186.

28 Cioffi, N., Ditaranto, N., Torsi, L. et al. (2005). Synthesis, analytical characterization and bioactivity of Ag and Cu nanoparticles embedded in poly-vinyl-methyl-ketone films. *Anal. Bioanal. Chem.* 382 (8): 1912–1918.

29 Pan, Y., Neuss, S., Leifert, A. et al. (2007). Size-dependent cytotoxicity of gold nanoparticles. *Small* 3 (11): 1941–1949.

30 Braydich-Stolle, L., Hussain, S., Schlager, J.J., and Hofmann, M.C. (2005). In vitro cytotoxicity of nanoparticles in mammalian germline stem cells. *Toxicol. Sci.* 88 (2): 412–419.

31 Hussain, S.M., Javorina, A.K., Schrand, A.M. et al. (2006). The interaction of manganese nanoparticles with PC-12 cells induces dopamine depletion. *Toxicol. Sci.* 92 (2): 456–463.

32 Chen, X. and Schluesener, H.J. (2008). Nanosilver: a nanoproduct in medical application. *Toxicol. Lett.* 176 (1): 1–12.

33 Hogstrand, C. and Wood, C.M. The toxicity of silver to marine fish. In: *Proceedings of the 4th International Conference on Transport, Fate and Effects of Silver in the Environment* (eds. A.W. Andren and T.W. Bober), 109–112. Lexington Kentucky, USA: University of Kentucky; McMaster University, Hamilton,Ontario,Canada.

34 Eisler, R. (1996). A review of silver hazards to plants and animals. In: *Proceedings of the 4th International Conference on Transport Fate and Effects of Silver in the Environment* (eds. A.W. Andren and T.W. Bober), 143–144. Madison, WI: University of Vasconia sea giant institute Madison.

35 Throback, I.N., Johansson, M., Rosenquist, M. et al. (2007). Silver (Ag^{+}) reduces denitrification and induces enrichment of novel *nirK* genotypes in soil. *FEMS Microbiol. Lett.* 270 (2): 189–194. Madison, WI: Sea Grant Institute.

36 Wood, C.M., Playle, R.C., and Hogstrand, C. (1999). Physiology and modelling of mechanisms of silver uptake and toxicity in fish. *Environ. Toxicol. Chem.* 1 (18): 71–83.

37 Ajayan, P.M., Schadler, L.S., and Braun, L.S. (2006). *Nanocomposite Science and Technology*. New York: Wiley.

38 Kim, J. and Bruggen, B.V. (2010). The use of nanoparticles in polymeric and ceramic membrane structures: review of manufacturing procedures and performance improvement for water treatment. *Environ. Pollut.* 158 (7): 2335–2349.

39 Rozenberg, B.A. and Tenne, R. (2008). Polymer-assisted fabrication of nanoparticles and nanocomposites. *Prog. Polym. Sci.* 33 (1): 40–112.

40 Pomogailo, A.D. (2005). Polymer sol-gel synthesis of hybrid nanocomposites. *Colloid J.* 67 (6): 658–677.

41 Alonso, A., Macanas, J., Davies, G.L. et al. (2011). Environmentally-safe polymer-metal nanocomposites with most favorable distribution of catalytically active and biocide nanoparticles. In: *Advances in Nanocomposite Technology* (eds. A. Alonso, J. Macanás, G.-L. Davies, et al.), 176–200. Rijeka, Croatia: Intech.

42 Lapied, E., Nahmani, J.Y., Moudilou, E. et al. (2011). Ecotoxicological effects of an aged TiO_2 nanocomposite measured as apoptosis in the anecic earthworm *Lumbricus terrestris* after exposure through water, food and soil. *Environ. Int.* 37 (6): 1105–1110.

43 Muraviev, D.N., Macanas, J., Farre, M. et al. (2006). Novel routes for inter-matrix synthesis and characterization of polymer stabilized metal nanoparticles for molecular recognition devices. *Sens. Actuators, B* 1181 (2): 408–417.

44 Savage, N. and Diallo, N. (2005). Nanomaterials and water purification: opportunities and challenges. *J. Nanopart. Res.* 7 (4–5): 331–342.

45 Adams, L.K., Lyon, D.Y., and Alvarez, P.J.J. (2006). Comparative ecotoxicity of nanoscale TiO_2, SiO_2, and ZnO water suspensions. *Water Res.* 40 (19): 3527–3532.

46 Lu, A.H., Salabas, E.L., and Schuth, F. (2007). Magnetic nanoparticles: synthesis, protection, functionalization, and application. *Angew. Chem. Int. Ed.* 46 (8): 1222–1244.

47 Laurent, S., Forge, D., Port, M. et al. (2008). Magnetic iron oxide nanoparticles: synthesis, stabilization, vectorization, physicochemical characterizations, and biological applications. *Chem. Rev.* 108 (6): 2064–2110.

48 Medyak, G.V., Shunkevich, A.A., Polikarpov, A.P. et al. (2001). Features of preparation and properties of FIBAN K-4 fibrous sorbents. *Russ. J. Appl. Chem.* 74 (10): 1658–1663.

49 Grillo, R., Abhilash, P.C., and Fraceto, L.F. (2016). Nanotechnology applied to bio-encapsulation of pesticides. *J. Nanosci. Nanotechnol.* 16: 1231–1234. https://doi.org/10.1166/jnn.2016.12332.

50 Jiang, S., Eltoukhy, A.A., Love, K.T. et al. (2013). Lipidoid-coated iron oxide nanoparticles for efficient DNA and siRNA delivery. *Nano Lett.* 13: 1059–1064. https://doi.org/10.1021/nl304287a.

51 Mishra, S. and Singh, H.B. (2016). Preparation of biomediated metal nanoparticles. Indian Patent Filed 201, 611, 003, 248.

52 De La Torre-Roche, R., Hawthorne, J., Deng, Y. et al. (2013). Multiwalled carbon nanotubes and C_{60} fullerenes differentially impact the accumulation of weathered pesticides in four agricultural plants. *Environ. Sci. Technol.* 47: 12539–12547. https://doi.org/10.1021/es4034809.

53 Parisi, C., Vigani, M., and Rodriguez-Cerezo, E. (2015). Agricultural nanotechnologies: what are the current possibilities? *Nano Today* 10: 124–127. https://doi.org/10.1016/j.bios.2015.11.086.

54 Mishra, S., Singh, A., Keswani, C. et al. (2015). Harnessing plant-microbe interactions for enhanced protection against phytopathogens. In: *Plant Microbes Symbiosis: Applied Facets* (ed. N.K. Arora), 111–125. New Delhi: Springer.

55 Benoit, R., Wilkinson, K.J., and Sauve, S. (2013). Partitioning of silver and chemical speciation of free Ag in soils amended with nanoparticles. *Chem. Cent. J.* 7: 75. https://doi.org/10.1186/1752-153X-7-75.

56 Wang, P., Menzies, N.W., Lombi, E. et al. (2013). Fate of ZnO nanoparticles in soils and cowpea (*Vigna unguiculata*). *Environ. Sci. Technol.* 47: 13822–13830. https://doi.org/10.1021/es403466p.

57 Kim, S., Kim, J., and Lee, I. (2011). Effects of Zn and ZnO nanoparticles and Zn^{2+} on soil enzyme activity and bioaccumulation of Zn in *Cucumis sativus*. *Chem. Ecol.* 27: 49–55. https://doi.org/10.1080/02757540.2010.529074.

58 Du, W., Sun, Y., Ji, R. et al. (2011). TiO_2 and ZnO nanoparticles negatively affect wheat growth and soil enzyme activities in agricultural soil. *J. Environ. Monit.* 13: 822–828. https://doi.org/10.1039/c0em00611d.

59 Zhao, L., Hernandez-Viezcas, J.A., Peralta-Videa, J.R. et al. (2013). ZnO nanoparticle fate in soil and zinc bioaccumulation in corn plants (*Zea mays*) influenced by alginate. *Environ. Sci. Processes Impacts* 15: 260–266. https://doi.org/10.1039/C2EM30610G.

60 Simonin, M. and Richaume, A. (2015). Impact of engineered nanoparticles on the activity, abundance, and diversity of soil microbial communities: a review. *Environ. Sci. Pollut. Res.* 22: 13710–13723. https://doi.org/10.1007/s11356-015-4171-x.

61 Ge, Y., Schimel, J.P., and Holden, P.A. (2012). Identification of soil bacteria susceptible to TiO_2 and ZnO nanoparticles. *Appl. Environ. Microbiol.* 78: 6749–6758. https://doi.org/10.1128/AEM.00941-12.

62 Shahrokh, S., Hosseinkhani, B., and Emtiazi, G. (2014). The impact of nanosilver on bacterial aerobic nitrate reductase. *J. Bioprocess. Biotechnol.* 4: 162. https://doi.org/10.4172/2155-9821.1000162.

63 VandeVoort, A.R. and Arai, Y. (2012). Effect of silver nanoparticles on soil denitrification kinetics. *Ind. Biotechnol.* 8: 358–364. https://doi.org/10.1089/ind.2012.0026.

64 Frenk, S., Ben-Moshe, T., Dror, I. et al. (2013). Effect of metal oxide nanoparticles on microbial community structure and function in two different soil types. *PLoS One* 8: e84441. https://doi.org/10.1371/journal.pone.0084441.

65 Shen, Z., Chen, Z., Hou, Z. et al. (2015). Ecotoxicological effect of zinc oxide nanoparticles on soil microorganisms. *Front. Environ. Sci. Eng.* 9: 912–918. https://doi.org/10.1007/s11783-015-0789-7.

66 Simonin, M., Guyonnet, J.P., Martins, J.M. et al. (2015). Influence of soil properties on the toxicity of TiO_2 nanoparticles on carbon mineralization and bacterial abundance. *J. Hazard. Mater.* 283: 529–535. https://doi.org/10.1016/j.jhazmat.2014.10.004.

67 Dietz, K.J. and Herth, S. (2011). Plant nanotoxicology. *Trends Plant Sci.* 16: 582–589. https://doi.org/10.1016/j.tplants.2011.08.003.

68 Zhang, P., Ma, Y., Zhang, Z. et al. (2012). Comparative toxicity of nanoparticulate/bulk Yb_2O_3 and $YbCl_3$ to cucumber (*Cucumis sativus*). *Environ. Sci. Technol.* 46: 1834–1841. https://doi.org/10.1021/es2027295.

69 Vannini, C., Domingo, G., Onelli, E. et al. (2014). Phytotoxic and genotoxic effects of silver nanoparticles exposure on germinating wheat seedlings. *J. Plant Physiol.* 171: 1142–1148. https://doi.org/10.1016/j.jplph.2014.05.002.

70 Asli, S. and Neumann, P.M. (2009). Colloidal suspensions of clay or titanium dioxide nanoparticles can inhibit leaf growth and transpiration via physical effects on root water transport. *Plant Cell Environ.* 32: 577–584. https://doi.org/10.1111/j.1365-3040.2009.01952.x.

71 Yang, F., Liu, C., Gao, F. et al. (2007). The improvement of spinach growth by nano-anatase TiO_2 treatment is related to nitrogen photoreduction. *Biol. Trace Elem. Res.* 119: 77–88. https://doi.org/10.1007/s12011-007-0046-4.

72 Song, G., Gao, Y., Wu, H. et al. (2012). Physiological effect of anatase TiO_2 nanoparticles on *Lemna minor*. *Environ. Toxicol. Chem.* 31: 2147–2152. https://doi.org/10.1002/etc.1933.

73 Mahajan, P., Dhoke, S.K., and Khanna, A.S. (2011). Effect of nano-ZnO particle suspension on growth of mung (*Vigna radiata*) and gram (*Cicer arietinum*) seedlings using plant agar method. *J. Nanotechnol.* 7: 696535. https://doi.org/10.1155/2011/696535.

74 Lee, W.M., An, Y.J., Yoon, H., and Kweon, H.S. (2008). Toxicity and bioavailability of copper nanoparticles to the terrestrial plants mung bean (*Phaseolus radiatus*) and wheat (*Triticum aestivum*): plant agar test for water-insoluble nanoparticles. *Environ. Toxicol. Chem.* 27: 1915–1921. https://doi.org/10.1897/07-481.1.

75 Stampoulis, D., Sinha, S.K., and White, J.C. (2009). Assay-dependent phytotoxicity of nanoparticles to plants. *Environ. Sci. Technol.* 43: 9473–9479. https://doi.org/10.1021/es901695c.

76 Shah, V. and Belozerova, I. (2009). Influence of metal nanoparticles on the soil microbial community and germination of lettuce seeds. *Water Air Soil Pollut.* 197: 143–148. https://doi.org/10.1007/s11270-008-9797-6.

77 Lin, D. and Xing, B. (2007). Phytotoxicity of nanoparticles: inhibition of seed germination and root growth. *Environ. Pollut.* 150: 243–250. https://doi.org/10.1016/j.envpol.2007.01.016.

78 Khodakovskaya, M.V., Kim, B., Kim, J.N. et al. (2013). Carbon nanotubes as plant growth regulators: effects on tomato growth, reproductive system, and soil microbial community. *Small* 9: 115–123. https://doi.org/10.1002/smll.201201225.

79 Zhu, H., Han, J., Xiao, J.Q., and Jin, Y. (2008). Uptake, translocation, and accumulation of manufactured iron oxide nanoparticles by pumpkin plants. *J. Environ. Monit.* 10 (6): 713–717.

80 Lee, W.-M., Kwak, J.I., and An, Y.-J. (2012). Effect of silver nanoparticles in crop plants *Phaseolus radiatus* and *Sorghum bicolor*: media effect on phytotoxicity. *Chemosphere* 86: 491–499. https://doi.org/10.1016/j.chemosphere.

81 Jośko, I. and Oleszczuk, P. (2013). Influence of soil type and environmental conditions on ZnO, TiO_2 and Ni nanoparticles phytotoxicity. *Chemosphere* 92: 91–99. https://doi.org/10.1016/j.chemosphere.2013.02.048.

82 Dimkpa, C.O., McLean, J.E., Britt, D.W., and Anderson, A.J. (2012). CuO and ZnO nanoparticles differently affect the secretion of fluorescent siderophores in the beneficial root colonizer, *Pseudomonas chlororaphis* O6. *Nanotoxicology* 6: 635–642. https://doi.org/10.3109/17435390.2011.598246.

83 Hänsch, M. and Emmerling, C. (2010). Effects of silver nanoparticles on the microbiota and enzyme activity in soil. *J. Plant Nutr. Soil Sci.* 173: 554–558. https://doi.org/10.1002/jpln.200900358.

84 Colman, B.P., Arnaout, C.L., Anciaux, S. et al. (2013). Low concentrations of silver nanoparticles in biosolids cause adverse ecosystem responses under realistic field scenario. *PLoS One* 8: e57189. https://doi.org/10.1371/journal.pone.0057189.

85 Vittori Antisari, L., Carbone, S., Gatti, A. et al. (2013). Toxicity of metal oxide (CeO_2, Fe_3O_4, SnO_2) engineered nanoparticles on soil microbial biomass and their distribution in soil. *Soil Biol. Biochem.* 60: 87–94. https://doi.org/10.1016/j.soilbio.2013.01.016.

86 Ge, Y., Priester, J.H., Van De Werfhorst, L.C. et al. (2013). Potential mechanisms and environmental controls of TiO_2 nanoparticle effects on soil bacterial communities. *Environ. Sci. Technol.* 47: 14411–14417. https://doi.org/10.1021/es403385c.

87 Kibbey, T. and Strevett, K. (2019). The effect of nanoparticles on soil and rhizosphere bacteria and plant growth in lettuce seedlings. *Chemosphere* 221: 703–707. https://doi.org/10.1016/j.chemosphere.2019.01.091.

88 Chen, H. and Yada, R. (2011). Nanotechnologies in agriculture: new tools for sustainable development. *Trends Food Sci. Technol.* 22: 585–594. https://doi.org/10.1002/ps.1732.

89 Wirth, S.M., Lowry, G.V., and Tilton, R.D. (2012). Natural organic matter alters biofilm tolerance to silver nanoparticles and dissolved silver. *Environ. Sci. Technol.* 46: 12687–12696. https://doi.org/10.1021/es301521p.

90 Calder, A.J., Dimkpa, C.O., McLean, J.E. et al. (2012). Soil components mitigate the antimicrobial effects of silver nanoparticles towards a beneficial soil bacterium, *Pseudomonas chlororaphis* O6. *Sci. Total Environ.* 429: 215–222. https://doi.org/10.1016/j.scitotenv.2012.04.049.

3

Nanotechnology and Disease Management

CHAPTER MENU

3.1 Recent Advancements in Plant Nanotechnology

In nanotechnology and nanoparticle research, significant progress has been made with considerable interest in the application of nanotechnology in agriculture and the potential impacts on the environment and on food products [1]. Nanomaterials can be toxic because of their catalytic properties, sizeable bioactive surface, and sizes that allow them to pass through the cell membranes and communicate with cellular structures and biomolecules. Possible phytotoxicity may be due to connections between nanomaterials and soil components that can alter the nutrient uptake of plants. Besides, nanomaterials can be taken up by plants via roots and shoots, and transported and accumulated inside plants in which biomolecules and cells interact [2]. The phytotoxicity studies of various nanomaterials (CeO_2, ZnO, TiO_2, NiO, CuO, Ag, Au, SiO_2, fullerenes, carbon nanotubes (CNTs), and graphene oxide) are described in the following subheadings:

Nano-Technological Intervention in Agricultural Productivity, First Edition.
Javid A. Parray, Mohammad Yaseen Mir and Nowsheen Shameem.

3.1.1 Cerium Oxide (CeO_2) NPs

Because of the durability of CeO_2 NPs, which tend to have minimal dissolution in soil and plant tissues, cerium oxide NPs are used as fuel additives and as sunscreens in cosmetics. Evaluating their relationship with plants, however, is still necessary. The effect of CeO_2-NPs has recently been assessed on the antioxidant protection mechanism and oxidative stress of germinating rice seeds [3]. The seeds grown in CeO_2 NP suspensions for 10 days showed substantial growth without any signs of toxicity. The research showed that low concentrations of CeO_2 NPs resulted in a significant reduction in root and shoot H_2O_2 material. In comparison, no substantial variance of H_2O_2 was observed at high concentrations, but only an alteration in enzymatic activity, resulting in membrane damage and photosynthetic stress in the shoots. This behaviour seems to be due to CeO_2 NPs' reactive oxygen species (ROS) scavenging ability, which increases with the decrease in concentration and size of NPs. Nevertheless, the real situation of NPs in the atmosphere and their contact with the organic matter of the soil that will influence the absorption of nanoparticles needs to be considered. In another related analysis, pumpkin and wheat were exposed to CeO_2 NPs in gum arabic and fulvic acid hydroponic systems. It was produced without any toxic response, as NPs translocated only into pumpkin shoots. The amount of NPs associated with roots as unadulterated > fulvic acid > gum arabic was affected by the organic matter but did not impact the factor of translocation. However, the addition of natural organic matter affected the amount of CeO_2 associated with the roots, as the root adsorption of particles, stabilized by organic matter, was significantly reduced. Fulvic acid and gum arabic delays agglomeration between particles but prevents root sorption of NPs and decreases the absorption of NPs into pumpkin shoots [4]. Morales et al. [5] evidenced as CeO_2 NPs can be accumulated in plants of *Coriandrum sativum* L. without modification. It was observed that plants produced longer roots and shoots, with higher biomass production, and a significant increase of catalase activity in shoots and ascorbate peroxidase in roots at a conc. of 125 mg/kg of CeO_2. Besides, it shows that CeO_2-NPs have fertilizing effects and assist plants to grow stronger. To evaluate the impact on seed quality and second-generation seedlings' growth, tomato plants were treated with low concentrations of CeO_2 NPs (10 mg/l). The results suggested that the second-generation seedlings, grown from the seeds of treated plants, were smaller and weaker; besides, compared to the control, they produced extensive root hairs. This study provides the first proof of the transgenerational effect of CeO_2 NPs on tomato plant production and growth [6].

3.1.2 Silver NPs

Silver (Ag) is the most manufactured nanomaterial and is found in different commercial products for its antimicrobial properties with inhibitory and bactericidal effects. However, they are oxidized in water, can complex with anions, and can finally be converted to heavy metals. Heavy metals can produce mutagenesis, carcinogenesis, and are toxic and hazardous because they can accumulate in the edible parts of vegetables. For this reason, plants have a pivotal role to play in the fate and transport of NPs in the environment through their uptake and accumulation. Shams et al. [7] suspended AgNPs (50 nm) and sprayed them into different parts of the cucumber plant to screen these modifications. It has been observed that, with an increase in the NP concentration, the massive silver metal and growth index have also increased in addition to the morphological characteristics of the plant. The cucumber fruit performance was significantly enhanced by increasing the concentration of AgNPs as the plant height and the number of leaves, while the treatments did not substantially affect the fruit pH and the percentage of dry fruit matter. The number, weight, and length of fruit per plant have increased considerably in plants sprayed with AgNPs. AgNPs can be used in the formulation of pesticides to improve integrated pest management and obtain a pathogen-free environment around the plant. In this way, plant and disease infections can be prevented, improving plant growth and crop production.

Also, the antifungal activity of AgNPs in plant assessments indicates that both silver ions and nanoparticles influence the formation of spores and the progression of pathogenic fungi. The efficacy of AgNPs is much more generous with preventive use, which may facilitate the direct contact of silver with spores and prevent their viability [8]. *Eruca sativa* was treated with different concentrations of AgNPs and $AgNO_3$ to understand the morphological and proteomic changes caused by Ag and AgNPs [9]. AgNPs and $AgNO_3$ cause different plant reactions, i.e. increased root elongation, induction of proteins in the jacalin lectin family, induction of two key enzymes involved in cysteine biosynthesis, and activation of ROS detoxification pathways. On the other hand, both treatments cause changes in proteins involved in sulfur metabolism and redox regulation. Only AgNPs cause an enhancement in plant growth and protein alteration in the endoplasmic reticulum and vacuole, implying these compartments as targets for NP action [9].

Oukarroum et al. [10] evaluated the toxic effect of AgNPs on the aquatic plant *Lemna gibba*, exposed to different concentrations of NPs, to determine whether the behaviour is the same as that of the other plants. After seven days of treatment, a reduction in the viability of the cellular plant was detected, which is correlated with intracellular ROS production, increased formation of ROS, and induced oxidative stress within *L. gibba* L., man. The plant cells were proportional to the exposure of the AgNPs. It seems that this stress was caused by the release of free Ag inside the plant cells. Studies on silver nanotoxicity in plants have been controversially centred on whether the cause of toxicity is either the nanosize or the shape of the particles or their release as ionic Ag^+. Ag ions are one of the most toxic forms for a wide range of microorganisms among metal pollutants. Ag ions lead to the development of a wealth of silver-based products in many bacterial applications [11]. The studies agree that silver nanotoxicity is positively concentration dependent and negatively dependent [12].

3.1.3 Titanium Dioxide (ToO_2) NPs

Titanium dioxide NPs are manufactured for cosmetics worldwide. In particular, they are used in sunblock products to protect the skin from UV light. Increased UV-B radiation causes direct and indirect effects, including degradation and conformation changes in DNA, proteins, photosynthesis alterations, and plant growth morphology. These particles are also used in antibacterial products and in wastewater to decompose the organic matter. Several studies show effects in animals (cytotoxicity, neurotoxicity, and DNA damage) with controversial results, and only a few studies analyse the impact of nano-TiO_2 on plants with opposite results [13, 14].

In addition to exerting toxic effects, nano-TiO_2 has excellent biological and optical properties. Gao et al. [15] investigated the development, at different concentrations, of nano-anatase-TiO_2 solutions on seedlings of *Ulmus elongata*. The effects of nano-biological TiO_2 depend on its structure and plant organisms. As a toxicity mechanism, TiO_2-NPs improved the synthesis of carbohydrates and lipids and, after foliar applications, decreased the net photosynthetic rate through non-stomatal regulation and disrupted seedling development, increased leaf chlorosis, and induced defoliation.

Because of wide surface area, high thermal conductivity, and high photocatalytic capacity that can inhibit bacterial growth, nano-anatase also has positive effects. Song et al. [16] researched the impacts on the germination and root elongation of seeds, seedlings, and the absorption and physiological responses of mature plants to prevent misleading findings on phytotoxicity or positive results of TiO_2-NPs. To counteract the precipitation of NPs and to recreate the environmental conditions, a hydroponic device was used. NPs absorbed into seed tissues did not show any toxicity in this situation.

Nano-TiO_2 does not affect the germination of seeds, enzyme function, and the quality of chlorophyll, even though plants consume NPs. It also has a positive impact on root elongation. Nanosizes will also become more detailed and not impact plants. When subjected to various treatments, plants did not show any substantial difference in total antioxidant ability, superoxide dismutase (SOD) activity, and physiological differences. Nano-TiO_2 does not seem toxic at higher concentrations, demonstrating that it will not always lead to higher absorption. In particular, these findings show that the toxicity of nano-TiO_2 at high concentrations is species dependent relative to the previous study and suggest that nano-TiO_2 has positive effects on plants. Nano-TiO_2 is difficult to accumulate in plants and does not show toxic effects in plants grown in the soil and hydroponic systems even though it penetrates the seed coat and enters its tissues. TiO_2 has been determined not to be toxic to plants.

Nano-TiO_2 was compared with bulk TiO_2 in another study [17] to study their properties and phytotoxic or stimulatory effects on seed germination and Fennel development. Fennel seeds have poor germination (54%) and were exposed to nano-TiO_2 (60 ppm) for 14 days with improved seed germination indices, while bulk TiO_2 (40 ppm) reduced shoot biomass and germination of seeds. Consequently, an improvement in germination time and an improvement in indexes such as germination value, vigour index, and mean daily germination were observed with nanoformulation using the same concentration, obtaining a substantial response time [18]. This research, therefore, measured the effectiveness of the treatments concerning the size of the nanoparticles. However, a survey on model organisms such as daphnias, algae, rotifers, and plants was performed by Clément et al. [19]. The results show that in the tests carried out, TiO_2 NPs with an anatase crystal structure are toxic but have facilitated the growth of roots because of their antimicrobial properties at high concentrations. Simultaneously, rutile is lipophilic and forms large aggregates with lower toxicity than anatase in an aqueous medium.

Because of other surface properties, these various structures have different reactivates, resulting in more severe anatase toxicity. It has been shown from previous TiO_2 studies that different NP plants, size, and crystallinity determine other positive or negative effects on plants and the environment. However, the findings are often contentious [20]. For this reason, the impact of different NPs on the same plants were studied by Song et al. [21] using the same approach to explicitly compare the findings to understand the beneficial or toxic effects of these NPs. Significant nano-TiO_2 treatment-induced particle agglomeration is the characteristics of NPs as water soluble or aggregation are important Nano TiO_2 treatment induced particle agglomeration, interrupting in such way their uptake and decreasing their effects on plants. In plant roots, fruits, and leaves, all NPs have been absorbed. Still, some papers have also documented positive results of nano-TiO_2 on plant growth, indicating the need of studies of the ecological effects on animal and human health and management strategies in the particular site in which they occur, to mitigate their impact and monitor their diffusion in the ecosystem [21].

3.1.4 Zinc Oxide (ZnO) NPs

In skincare products such as toothpaste, sunscreens, soaps and shampoos, pigments and medicines, and cosmetics with a wide distribution in the environment, nanoparticles absorbed by various plant species can produce positive and negative effects, in particular (ZnO) NPs. The phytotoxicity of ZnO NPs was tested on *Fagopyrum esculentum* [22], a medicinal plant. Because it is known as a bioindicator of environmental conditions, it is easy to develop and respond to environmental stress. Besides, this plant is a hyperaccumulator of metals and can translocate metals from roots to shoots. These plants were cultivated in hydroponic cultures, and the presence of NPs was observed in root cells by scanning electron microscopy (SEM) and transmission electron microscopy (TEM) analyses. Using NPs instead of microparticles, substantial biomass reduction, stunted growth, and a higher translocation of Zn in plants were observed. They further stated that at high NP concentrations, the amount of GSH increased because NPs have activated the antioxidant defence mechanism. Catalase (CAT) activity has also risen at the same time. A large amount of ROS has been produced at high doses of NPs, and CAT loses the scavenging capacity, causing oxidative damage. This study shows that this type of nanomaterial appears only to have harmful effects on the environment. Nevertheless, the toxicity limit was further examined by Zhao et al. [23]. The maize plants were cultivated in modified soil with ZnO NPs and sodium alginate for one month. Alginic acid is a part of natural organic matter, and its sodium salt has been used as a capping agent to stabilize NPs in suspension. Previous studies have shown that ZnO NPs have detrimental effects on soil and plants, but they seem to depend on the soil's quality and the species of plants. Without a decrease in biomass production, plants treated with a definite concentration of ZnO-NPs plus alginate had Zn in their tissues. It may be because alginate raises the amount of chlorophyll *a*, helping to preserve photosynthesis. At the same time, alginate decreased the activity of stress peroxidase and catalase enzymes, caused damage to the protection system, increased the release of Zn ions into the soil solution, and facilitated bioaccumulation in the tissues of maize plants, resulting in more than Zn in the aerial portions of the plants. This shows that alginate can modulate the fate of ZnO NPs in soil and plants, a natural component of ecosystems, reducing its toxic effect on plants.

3.1.5 Cupric Oxide (CuO) NPs

Because of their greater surface area and smaller size than microparticles, NPs are bioreactive, and it is essential to consider their effect on ecosystems. The phytotoxic and genotoxic effects of two forms of ZnO and CuO NPs on buckwheat seedlings were compared by Lee et al. [24] to estimate the toxic impact of NP-released ions on biomass and inhibition of root growth. ZnO NPs are used for environmental remediation in chemotherapeutic medicines, personal care products, and coating catalysts and can pass into the atmosphere and be consumed by plants. These NPs and significant root morphology changes such as inhibition of root length, a decrease in root development, and localization of NPs within the root epidermis cells affected buckwheat seedling growth in particular. Various studies have shown the phytotoxicity of CuO and ZnO NPs for different plants and Zn's translocation to plant shoots compared to Cu. The activity of CuO NPs in the climate, particularly in wheat plants grown in sand treated with these NPs, was assessed by Dimkpa et al. [25]. CuO NPs are phytotoxic to the wheat root, resulting in a reduction in volume. Their involvement in the root surface was noticed when the sand was adjusted with these NPs and then translocated into plant aerial portions to form complexes with different ligands to understand the fate of NPs from soil to plants, confirming the phytotoxicity of CuO NPs [26, 27].

3.1.6 Gold NPs (GNPs)

GNPs are among the most synthesized NPs since the sixteenth century and have been used mainly for medical purposes. Even, there are very few studies recording their effects on plant physiology, growth, and metabolism. The cost-effective method for preparing GNPs for various agricultural applications has been demonstrated by Kumar et al. [28]. *Syzygium cumini* leaf extract was synthesized by AuNPs and exposure experiments were performed on *Arabidopsis thaliana*. Exposure to AuNPs was found to improve *A. thaliana* compared to control and increase the germination of seeds, the activity of free radical scavenging, and vegetative growth. The fresh biomass, seed germination rate, and relative water content of *A. thaliana* substantially increased with gold nanoparticles (GNP) exposure relative to control with the potential for free radical scavenging. This tends to be due to increased seed ability for water uptake and increased seed germination and new biomass rate. APx, CAT, GR, and SOD activity is essential for the plant protection system against ROS. It has been enhanced in dose-dependent seedlings exposed to GNPs acting in plant antioxidant defence. Compared to regulation, GNP exposure was found to increase the root and shoot length and alter levels of expression of microRNAs involved in controlling various morphological, physiological, and metabolic processes in plants.

3.1.7 Carbon Nanotubes

CNTs are interesting for their strength, nanosize, and physicochemical characteristics. Based on their ability to cross the cell membrane of mammalian cells that function as drug delivery vehicles, they have gained interest as potential molecular transporters; they can be released into the environment in this way. CNTs have been deposited on the seed surface by air spray or added to the plant growth medium, and it has been demonstrated that they can penetrate the seed coats and activate germination [29]. Multiwalled carbon nanotubes or MWCNTs were applied to the agar medium in particular, and seed germination and growth activation were observed in three crops: barley, soybean, and maize.

When MWCNTs were deposited on seed surfaces, seed germination was enabled. The expression of genes encoding various water channel proteins (aquaporins) was increased in treated seeds, as evidenced by real-time reverse transcription–polymerase chain reaction (RT-PCR). These findings demonstrate the positive effects of CNTs on seed germination and growth in various crop species, demonstrating their ability as regulators of germination and plant growth. Functionalized CNTs' power to penetrate the plant cell wall and cause specific organelle changes is shown in a further study [30]. Their tobacco and seed coat cell membrane penetration properties stimulate germination and encourage growth in tobacco and tomato plants.

Depending on their diameters, CNTs penetrate the plant cell wall in two ways: if the average diameter of CNTs is below the diameter of a plant cell wall, they can enter the apoplast, but chemical methods such as ultrasonic-assisted chemical oxidative cutting can help the tubes move through the cell wall structure, and if the diameter of the CNT exceeds the 5 nm pores of the plant cell wall. The immobilization of cellulose molecules on the walls and tips of CNTs that aid their penetration is another recent penetration strategy. Therefore, the subcellular position of CNTs depends on their duration and the extent of the adsorption of the tag on their surface. This is essential for the delivery to specific subcellular organelles of substances. Thus, they can be used as wood materials or as gene delivery vectors to plant cells to strengthen fibres. In this way, advantages in agricultural biology and plant biotechnology can be achieved at the plant material level, for the whole plant, and the plant's cells and molecules [31].

3.1.8 Nickel Oxide NPs

In various formulations, nickel oxide NPs are used as antibacterial materials. Faisal et al. [32] investigated the phytotoxicity induced by NiO-NPs in tomato seedling roots and their root cell translocation and changes in cell organelles that induced apoptosis in tomato root cells, triggering the release of mitochondrial caspase-3 proteases, leading to dysfunction of mitochondria. This study also demonstrates their ability via the apoptotic and necrotic pathway to generate cell death.

3.2 Nanotechnology: Role in Plant-Parasitic Control

Nanotechnology manipulates matter on the scale of atoms and molecules with sizes ranging from 0.1 to 100 nm [33]. In medicine, materials and electronics research in nanoscience has increased significantly [34, 35]. A new window has opened up the possible applications of the nanoworld in different fields, such as agriculture and food systems, and its use in developing countries is genuinely among the enabling technologies [35, 36]. The goals to be integrated into agriculture and food systems are delivery systems, nanosensors, nanomaterials, etc. Some of the possible aspects of nanotechnology are new devices and techniques for molecular and rapid identification and treatment of diseases, enhancing plants' ability to consume nutrients, and delivering specific substances to specific locations. In the production and formulation of agrochemicals, nanodevices such as liposomes or nanocapsules are assimilated, and their futuristic function is discussed in the same way as in [37, 38].

3.2.1 Nanocapsules: Liposomes and Polymers

For devices consisting of a thin external shell with space within, the term nanocapsule is usually intended. Encapsulation and controlled release methods are swapping the use of drugs in medicine and of pesticides and fertilizers in agriculture. This technology enables formulations to be made in which the active compound within a shell is enclosed by the nanocapsules, generally composed of a polymer. The shell protects the chemical against harm from external agents. It helps boost solubility and penetration through tissues depending on the external environments, making them more desirable for managed unloading of the charge once they enter destinations, e.g. alkaline environment of some insects' stomach. With many uses in agriculture and food technology, liposomes are also being used in pharmaceuticals and cosmetics. They are spherical bilayer vesicles formed by the dispersion of aqueous solvent polar lipids, with sizes ranging from the nano- to micrometre range. Their properties depend on external and internal variables such as temperature, pH, ionic strength, concentration, and composition of phospholipids. Liposomes have been mainly used as model membranes of plant organelles in agriculture, leading to studies on plant ageing, drying, and freezing tolerance and membrane effects of toxins and pesticides [39, 40]. However, they are mostly used for flavour compounds, protecting substances within goods before they are released to boost food safety, such as antimicrobials [41].

The essential components for nanocapsules [42] have recently become polymers. Powerful covalent bonds keep polymeric capsules together, making them more robust and stable than liposomes, also in the dry form [43]. Using polymers that can be functionalized with various substances such as chitosan and chitin derivatives [44], the nanocapsules' sizes and shapes can be regulated more effectively. They are polysaccharides composed of linear β-1,4-configuration glucose residues resulting in rigid and unbranched structures with good mucoadhesive properties, facilitating the transport of drugs via cellular membranes [45, 46], which determine potential uses in the agrochemical field. In addition to inducing defensive responses in plants [47, 48], chitosans are best known in agriculture for their antimicrobial properties such as antifungal and antibacterial activities [49, 50].

3.2.1.1 Potential Uses in Controlling Parasitic Weeds

Nanoencapsulation may be used to overcome the phytotoxicity problems of crop herbicides used against parasites [51]. Tubercles of the broomrape (*Orobanche* spp.) serve as powerful sinks for nutrients and water, as well as for systemic herbicides [52]. The herbicide will be released at the destination as soon as the nanocapsules reach the parasitic weed, i.e. lower doses of herbicides such as glyphosate or imidazolinones are required because they are either not degraded by the crop or accumulate in the parasitic weed because of the sinking effect and are activated within the parasitic weed under certain conditions. To monitor *Orobanche* spp., sulfonylurea herbicides are added to the soil [53]. If the herbicide was encapsulated and slowly released into the soil during the crop season, better control could be achieved with one single application and lower herbicide rates [54].

Because of their amphiphilic composition, the liposomes and lipid vesicles can cross quickly through the plant cuticle [55]. To modify the surface characteristics as required, polymeric nanocapsules can be functionalized with various substances. For instance, adjuvants such as oil concentrates or surfactants may be bound directly to the shell of the nanocapsule, allowing greater penetration through the cuticle. It may also be possible to develop and separately encapsulate an agrochemical containing various substances for various functions such as plant protection, fertilization, hormones, etc., to avoid interactions and degradation between them. Nanoencapsulation can also make it possible to battle herbicides against parasitic weeds that are usually not systemically applicable, increasing every single treatment's effectiveness and reducing the lethal amount to be delivered. Within nanocapsules, germination stimulants [56] may be produced in the soil, protecting them against degradation. It was also possible to systemically apply natural metabolites, such as mycotoxins [57]. For the absorption of chemical signals released by the host, such as germination stimulants or other causes, nanocapsules delivered into the rhizosphere may also be used [58].

3.3 Abiotic Stress-Tolerant Transgenic Crops and Nanotechnology

The various transgenes were studied to increase the resistance of crops to salt, drought, and variable temperatures. Similarly, many genes have made it possible to grow transgenic rice against the extremes of drought, salt, flooding, and temperature. Many model plant species such as Arabidopsis and tobacco and crops such as tomato and wheat were genetically transformed for different abiotic stresses. Specifically, specific reports indicate that plants have been successfully developed by transgenic methods with improved abiotic stress tolerance [59].

Three essential responses [60] control the effectiveness of transgenic plants' development in principle: (i) availability of genes effectively, (ii) pertinent techniques, and (iii) transgene expression in genetic engineering studies, the transition of genes is often a spontaneous occurrence. Experiments have shown that the trans-hosts' level of expression of the transcripts/proteins from the transgenesis is very variable.

Therefore, a great deal of effort is required to select the progenies that express the trans-protein optimally. The progenies need to be adequately screened to establish transgenic forms with the optimum levels of gene expression. Gene transfer technologies need to be designed for this feedback acceptably for practical biotechnological applications.

3.3.1 Nanotechnology in Gene Transfer Experiments

In the physical, biological, agricultural, and medical sciences, nanotechnology concepts have been widely applied in a highly beneficial way [61]. In addition to addressing the question of rice husk disposal, the use of nanotechnology to manufacture nanosilica from rice husk has also been useful in manufacturing substances such as glass and concrete [62].

Using nanotechnology tools [63], delivery systems for pests, nutrients, and plant hormones have been developed. To enhance plant disease resistance, nutrient usage, and plant growth, nanotechnology enables the controlled release of agrochemicals, allowing direct delivery of different macromolecules at pre-conceived sites [64]. In terms of use and safe handling of pesticides, nanoencapsulation improves efficiency [64]. Torney et al. [65] For their onward journey to segregated plant cells and intact leaves, a honeycomb mesoporous silica nanoparticle (MSN) system with 3 nm pores could serve as an efficient conduit for

DNA chemicals. The researchers demonstrated transient expression in tobacco mesophyll protoplasts using MSN particles coated with a plasmid expressing a green fluorescent protein under the CaMV35S promoter. After the DNA-coated MSN particles' mesopores were capped by gold nanoparticles, intact plant cells of tobacco cotyledons were also successfully transformed using the gene shot.

This method's efficacy for producing stable transgenic plants was developed by raising the callus of immature maize embryos bombarded with GFP. The researchers subsequently made transgenic tobacco plants containing an inducible promoter-driven GFP gene, which could be chemically activated by the addition of β-oestradiol. Transgenic plantlets were bombarded with β-oestradiol-loaded MSN particles that had their pores capped by gold nanoparticles to avoid leaching. The presence of dithiothreitol (DTT) in the germination medium as an uncapping agent allowed the controlled release of the chemical inducer, followed by GFP transgene expression. Gold nanoparticles can prevent molecules from moving away. Undoing the capping in the presence of DTT resulted in the regulated release of gene expression in plants by chemicals [65].

3.4 Plant Pathogens and Nanoparticle Biosynthesis

Physical and chemical techniques have been used for the production of nanoparticles [66]. These techniques are costly and use toxic chemicals to synthesize UV radiation, laser ablation, and aerosol spray nanoparticles [67]. The physical approach limits synthesis and requires higher energy utilization, and costs are very high. However, chemical approaches are low cost but require toxic solvents and traces of chemical contamination by the generation of toxic by-products. The biological process for the inclusion of bacteria, fungi, yeast, and viruses provides high-yield, low-cost, and environmentally sustainable nanoparticle synthesis [68]. The synthesis of nanoparticles from plant pathogens is also being explored [67].

The microbial synthesis of nanoparticles is considered an exciting research field for potential use [66, 69] with rich microbial biodiversity; simple cellular, biochemical, or molecular microbial cultivation; synthesis; and improvement of nanoparticles' properties [67]. Biotechnological processes, such as bioleaching, biomineralization, biocorrosion and bioremediation, and microbial and metal interaction, are well recognized [67, 70]. The properties of microbe-manufactured nanostructured mineral crystals and metallic nanoparticles are similar to those of chemically synthesized nanomaterials [70]. Many unicellular and multicellular species have inorganic materials that are either intracellular or extracellular [68, 69]. To a certain degree, the formation of these inorganic materials can be manipulated by regulating the culture parameters [71] for shape and scale. The mechanism of nanomaterial intracellular and extracellular development varies across various microbes [69, 72]. Intracellular processes require beneficial transportation of metal ions into the cell wall by contacting the cell wall's negative ions. Besides, the metal ion is converted into nanoparticles by cell wall enzymes, and then, these nanoparticles disperse through the bacteria's cell wall. Nitrate reductase-mediated synthesis in the presence of enzyme nitrate, which helps bio-reduce metal ions into nanoparticles, is the extracellular production of nanoparticles in the case of fungi [72]. The various bio-mediators used and their characteristics are addressed as follows.

3.4.1 Bacteria-Mediated Biosynthesis

The critical metals must percolate into the cytoplasm through the bacterial wall and be transferred back for extracellular extraction through wall meshwork. The peptidoglycan cell wall composition provides poly-anions for stoichiometric interaction between metal and cell wall reactive chemical groups, accompanied by inorganic metal deposition. Cell walls with multiple possible metal binding sites can be affected by chemical reactions for particular groups, such as amines and carboxyl groups, which transform positive to negative charges. Chemical alteration of peptidoglycan of *Bacillus subtilis* has resulted in a transformation of the glutamic acid carboxyl group for fast metal penetration with deep metal deposition [73]. Once inside the cell, the metal deposition can range from 20 000 to 40 000 fold over the extracellular concentration and give the dipole moment to the bacteria for orientation toward the geomagnetic field. The crystalline and non-crystalline phases of the particles are often influenced by the intra- and extracellular cellular environment and species with a specific bacterial morphology. The intracellular and extracellular method of synthesizing nanoparticles is shown in Figure 3.1.

In the presence of iron oxide hydrates as a precursor [74], the mechanism of Fe_3O_4 crystal formation occurs within cells during the crystal growth mechanism. For example, *Klebsiella pneumoniae* was able to form nanometre-sized CdS particles on the cell surface in the presence of Cd^{++} ions in the growth medium. The bacterial cell wall's surface characteristic reflects the deposition of discreet CdS or bio-CdS quantum dots for possible biological membrane protection against corrosion for potential bio-semiconductor applications [75]. Using *Pseudomonas stutzeri*, the cermet nanoparticle was formed between the cell wall and the plasma membrane by isolated strains from the silver mine. Biologically synthesized nanoparticles have complementary properties to traditional thin-film coating

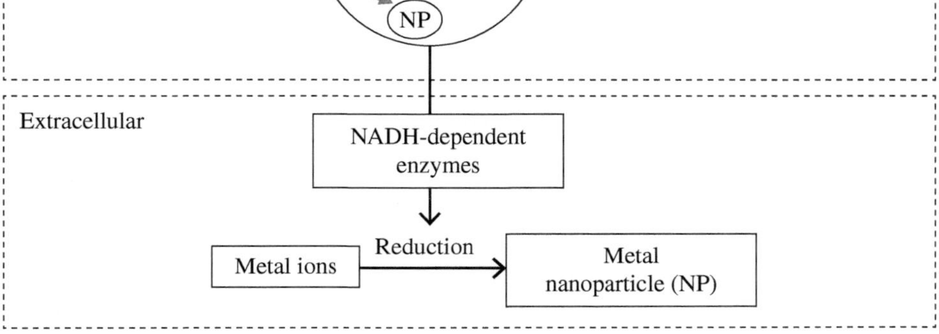

Figure 3.1 Nanoparticle biosynthesis via intracellular and extracellular methods. Source: Das et al. [73]. © 2017, Springer Nature.

technologies because of their low cost and relatively simple process. These substances have useful technological applications in optical filters and can also be used effectively to convert photothermal solar energy [76, 77]. The silver resistance of *P. stutzeri* allows the cellular efflux pumping mechanism to bind cell surface proteins by metal flux and metal binding to form and absorb precipitated metals [77]. The resistance of bacteria to metal ions and metal nanoparticles' formation depends on physical and chemical growth parameters, such as particle size, pH, temperature, growth time, growth media composition, and light and dark growth conditions. Biosynthesis of metal-containing bacteria nanoparticles gives regulated optical and electrical properties for potential future use [78]. Bacterial nanomaterials can be entirely harvested and fully used in analytical chemistry, metal ion recovery, rheumatology, and drug delivery in the medical field [75]. Green nanofactories offer the biological synthesis of spherical platinum nanoparticles using living and non-living organisms, a feasible alternative to conventional chemical methods [79, 80]. Surgical tools made of titanium react to a human serum that causes cancer cells to be produced. Nevertheless, the potential nano-titanium for future cancer chemotherapy and gene delivery can be emphasized with the eco-friendly biological approach using bacteria [79, 80]. Table 3.1 illustrates how nanoparticles are processed at various sites with different types and sizes of bacteria.

3.4.2 Fungal Mediated Biosynthesis

Nanoparticles with nanoscale dimensions and a more tolerable monodispersion can be synthesized with fungi than bacterial synthesis. Fungi have potential strategies for extracellular nanoparticle synthesis for better commercial feasibility. *Fusarium oxysporum* releases reducing agents into the solution for the formation of gold NPs in the presence of aqueous $AuCl_4$ ions with an NADH-enzyme-mediated response. Because of the protein binding through cysteine and lysine residues' binding, the synthesized nanoparticles

Table 3.1 Description of bacterial species for the synthesis metallic nanoparticles.

Microorganisms	NPs	Morphology	References
B. subtilis	Gold	Intracellular wall	[73]
Bacillus licheniformis	Silver	Extracellular	[81]
Bacillus cereus	Silver	Intracellular	[82]
Aquaspirillum magnetotacticum	Ferrous oxide	Intracellular	[74]
K. pneumoniae	Cadmium sulfide	On cell surface	[75]
P. stutzeri AG259	Silver, silver sulfide	Periplasmic space	[76]
Lactobacillus sp.	Gold, silver,	Intracellular	[77]
Corynebacterium sp.	Silver	Within the cell wall	[78]
Lactobacillus sp.	Titanium	Intracellular	[83]
Pediastrum boryanum	Platinum	Intracellular	[84]
Escherichia coli	Gold	Extracellular	[85]
E. coli	Cadmium	Extracellular	[86]

Table 3.2 Description of fungal species used for the synthesis of different metallic nanoparticles.

Fungal species	Nanoparticle	Localization/morphology	References
Fusarium exosporum	Au	Extracellular	[87]
Fusarium oxysporum	Zr	Extracellular	[88]
Fusarium oxysporum	$BaTiO_3$	Extracellular	[88]
Fusarium oxysporum	Si, Ti	Extracellular	[89]
Fusarium oxysporum	Au–Ag	Extracellular	[92]
Fusarium oxysporum	Pt	Extracellular	[93]
Verticillium luteoalbum	Au	Extracellular	[70]
Aspergillus flavus	Ag	Extracellular	[90]
Coriolus versicolor	Ag	Intracellular and extracellular	[91]
Guignardia mangiferae	$AgNO_3$	Extracellular	[94]
Penicillium oxalicum GRS-1	$AgNO_3$	Intracellular	[95]
Duddingtonia flagrans	AgNPs	Extracellular	[96]
Trichoderma harzianum	$AgNO_3$	Extracellular/intracellular	[97]

show long-term stability in the solution. Because of this property, immobilization for optoelectronic and nonlinear optical application on matrices or thin films is possible [87].

In comparison with other fungus species, *F. oxysporum* can hydrolyze metals that have never been encountered, and crystalline zirconia NPs are formed in the presence of K_2ZrF_6 aqueous solution. In the presence of aqueous anionic complexes of Si and Ti, the fungus can also synthesize silica and titanium nanoparticles, respectively. The fungal biological system has an eco-friendly and energy-conserving regenerative capacity for large-scale synthesis of metal nanoparticles for potential commercial viability [88, 89].

To manipulate the size and shape of gold nanoparticles by altering the pH and temperature during growing conditions [70], various bacteria, fungi, and yeasts isolated from soil- and metal-rich dump samples have been used. *Aspergillus flavus* showed the synthesis of monodispersed silver nanoparticles with an average particle size of 8.92 nm. The ‘sil’ gene was also present in plasmids to reduce silver ions for large-scale production [90]. The silver nanoparticles synthesized by the fungus as stabilizing agents in the presence of glucose had potential applications for living cells as water-soluble metallic catalysts [91]. Table 3.2 presents the development of nanoparticles using various fungi with different shapes and sizes.

3.5 Nanomaterial and Plant Protection Against Pests and Pathogens

Worldwide plant pathologists work toward a productive resolution for food protection of food and agricultural productivity from pathogens. For crop protection, it is therefore essential to combine nano-methods with the existing ones. Nanotechnology endorses the synthesis of NPs with contractible shape and size from fungi, which is futuristic hope. Possible applications of myconanotechnology deliver exhilarating impressions of transformation in farming and inspire microbiologists to establish food safety techniques using green chemistry [98]. In a particular timeframe, nanotechnology has the potential to reduce crop losses [99]. Nanotechnology applications offer valuable opportunities to monitor their release with too much expertise and unsafe treatment of fungicides, herbicides, and fertilizers [100].

There are currently efforts to develop disease control strategies and are relatively less dangerous with minimal use of synthetic fungicides. There has only been little exposure to the antifungal activity of silver NPs [101]. *Botrytis cinerea* induces that grey mould [102] has the high inhibitory ability in silver NPs for fungi. The Kirby–Bauer disc diffusion method has been used to assess the effect of fluconazole and silver NPs on three fungal pathogens, *Phoma glomerata*, *Phoma herbarum*, and *Fusarium semitectum*, for antifungal activity [103]. Ag_2S nanocrystals demonstrated antifungal activity against *Aspergillus niger* [104] on amorphous silica particles. The antifungal effects of silver NPs were evaluated by Min et al. [105], especially on sclerotia-forming phytopathogenic fungi. In South Korea, the antifungal activity of silver NPs against filamentous ambrosia fungi was assessed [83]. To exhibit antifungal activity against *F. oxysporum*, the effect of silver NPs on plant pathogenic spores of *Fusarium culmorum* [84] was studied [106] and the number of germinating fragments and sprout length decreased relative to control. In comparison to synthetic fungicides, the new synthesized silver NPs could be non-toxic to humans and animals. On *Colletotrichum gloeosporioides*, responsible for causing fruit anthracnose, the antifungal activity of the silver NPs was evaluated [107, 108].

Intelligent agricultural delivery systems may have timely, spatially targeted, automatic, remotely controlled, preprogrammed, or multifunctional characteristics to avoid biological barriers to successful targeting. González-Melendi et al. [109] work on in vitro treated zucchini (*Cucurbita pepo*) plants with carbon-coated Fe·NPs improve advanced delivery systems in the field. González-Melendi et al. [109] were the first to record within the entire plant the penetration and transport of NPs. Plants respond to different stress conditions and fight them by physiological changes [110], i.e. systemic defence induction [111]. By harnessing this indirect stimulus, Wang et al. [112] produced a sensitive electrochemical sensor, and the use of gold electrodes with copper NPs was demonstrated to check salicylic acid and pathogen detection levels. In a simple, cost-effective manner, nano-dispersed formulations can be prepared and are ideal for developing new types of fungicidal materials. Syngenta's Banner MAXX™ microemulsion concentrate formulations, offering excellent tank mix compatibility and stability, are among the most commonly used nanoformulations for treating diseases in turf and ornamental plants, soil, and seeds [113]. The resistance in plants resolves the economic loss issue by handling the above agents. Plant transformation by nanoparticles can boost the genetic engineering of plants further. Specific problems may mostly be solved in the field of plant defence, pathogen identification, and deciphering plant–pathogen interactions by using the nanoparticles in agriculture and new methods for the management of plant diseases [114]. The use of resistance genes for creating resistant variables in plant cells, for example, would reduce the costs of agrochemicals. Nano-phytopathology is used to comprehend the interactions between plants and pathogenic agents and provide new plant protection methods.

3.6 Future Perspectives

By overcoming the use of pesticides such as insecticides, fungicides, and herbicides, nanotechnology has entered the field of pesticides and can pest control. Atomically engineered plants can substitute modern nanoscale methods for integrating and harnessing genes for genetically engineered plants [115]. Pesticides may be packed to kill unwanted pests more accurately. In compressing the material to the nanometre scale, nanotechnologies' crux continues by growing the size [116]. The use of silicon because of its non-residual properties is incorporated into the field management of pests and diseases. Because NPs can control pest populations, additional studies are needed concerning chemical pesticides to assess their viability and environmental effects. The introduction of genetic modifications, policy and economic improvements, and mitigation strategies will contribute to future adjustments to the impact of climate change on plant diseases. The adoption of new crop plants, which reduce GHG production through lower nitrous oxide release, can reduce GHG quantities. Nano-article-driven plant genetic engineering and nanosensors in agriculture are boundary areas where crop production can be improved by preventing farmers' pesticide and disease incidents. Some of the NPs in the field of control of plant diseases are carbon, silver, silica, and alumina nanoforms. The active carrier ingredient of pesticides, host protective chemicals caused by the target pathogens, can be used in the same organic pesticides mode. Because of its microscopic scale, nanoparticles can hit/target virus particles and open up a new herbal virus control area. To mitigate the impact of climate change, an analysis of the ecology and genetic interaction would allow for a deeper understanding of the genetic constraints for pathogen and plant adaptation.

References

1 Mura, S., Seddaiu, G., Bacchini, F. et al. (2013). Advances of nanotechnology agro-environmental studies. *Italian J. Agron.* 8: 127–140. https://doi.org/10.4081/ija.2013.e18, eISSN2039-6805.

2 Masarovicova, E. and Kralova, K. (2013). Metal nanoparticles and plants. *Ecol. Chem. Eng. S* 20 (1): 9–22.

3 Rico, C.M., Hong, J., Morales, M.I. et al. (2013). Effect of cerium oxide nanoparticles on rice: a study involving the antioxidant defence system and in vivo fluorescence imaging. *Environ. Sci. Technol.* 47 (11): 5635–5642.

4 Schwabe, F., Schulin, R., Limbach, L.K. et al. (2013). Influence of two types of organic matter on interaction of CeO_2 nanoparticles with plants in hydroponic culture. *Chemosphere* 91 (4): 512–520. https://doi.org/10.1016/j.chemosphere.2012.12.025.

5 Morales, M.I., Rico, C.M., Hernandez-Viezcas, J.A. et al. (2013). Toxicity assessment of cerium oxide nanoparticles in cilantro (*Coriandrum sativum* L.) plants grown in organic soil. *J. Agric. Food Chem.* 61 (26): 6224–6230. https://doi.org/10.1021/jf401628v.

6 Wang, Q., Ebbs, S.D., Chen, Y., and Ma, X. (2013). Trans-generational impact of cerium oxide nanoparticles on tomato plants. *Metallomics* 5 (6): 753–759. https://doi.org/10.1039/c3mt00033h.

7 Shams, G., Ranjbar, M., and Amiri, A. (2013). Effect of silver nanoparticles on concentration of silver heavy element and growth indexes in cucumber (*Cucumis sativus* L. *negeen*). *J. Nanopart. Res.* 15 (5): 1630–1642.

8 Ouda, S.M. (2014). Antifungal activity of silver and copper nanoparticles on two plant pathogens, *Alternaria alternata* and *Botrytis cinerea*. *Res. J. Microbiol.* 9: 34–42. https://doi.org/10.3923/jm.2014.34.42.

9 Vannini, C., Domingo, G., Onelli, E. et al. (2013). Morphological and proteomic responses of *Eruca sativa* exposed to silver nanoparticles or silver nitrate. *PLoS One* 8 (7): e68752. https://doi.org/10.1371/journal.pone.0068752.

10 Oukarroum, A., Barhoumi, L., Pirastru, L., and Dewez, D. (2013). Silver nanoparticle toxicity effect on growth and cellular viability of the aquatic plant *Lemna gibba*. *Environ. Toxicol. Chem.* 32 (4): 902–907. https://doi.org/10.1002/etc.2131.

11 Boenigk, J., Beisser, D., Zimmermann, S. et al. (2014). Effects of silver nitrate and silver nanoparticles on a planktonic community: general trends after short-term exposure. *PLoS One* 9: e95340. https://doi.org/10.1371/journal.pone.00953.

12 Geisler-Lee, J., Brooks, M., Gerfen, J. et al. (2014). Reproductive toxicity and life history study of silver nanoparticle effect, uptake and transport in *Arabidopsis thaliana*. *Nanomaterials* 4: 301–318. https://doi.org/10.3390/nano4020301.

13 Ruffini, M., Giorgetti, L., Geri, C., and Cremonini, R. (2011). The effects of nano-TiO_2 on seed germination, development and mitosis of root tip cells of *Vicia narbonensis* L. and *Zea mays* L. *J. Nanopart. Res.* 1 (6): 2443–2449. https://doi.org/10.1007/s11051-010-0135-8.

14 Zheng, L., Hong, F., Lu, S., and Liu, C. (2005). Effect of nano-TiO_2 on strength of naturally aged seeds and growth of spinach. *Biol. Trace Elem. Res.* 104 (1): 83–91. https://doi.org/10.1385/bter:104:1:083.

15 Gao, J., Xu, G., Qian, H. et al. (2013). Effects of nano-TiO_2 on photosynthetic characteristics of *Ulmus elongata* seedlings. *Environ. Pollut.* 176: 63–70. https://doi.org/10.1016/j.envpol.2013.01.027.

16 Song, U., Shin, M., Lee, G. et al. (2013). Functional analysis of TiO_2 nanoparticle toxicity in three plant species. *Biol. Trace Elem. Res.* 155 (1): 93–103. https://doi.org/10.1007/s12011-013-9765-x.

17 Feizi, H., Kamali, M., Jafari, L., and Rezvani Moghaddam, P. (2013). Phytotoxicity and stimulatory impacts of nanosized and bulk titanium dioxide on Fennel (*Foeniculum vulgare* Mill). *Chemosphere* 91 (4): 506–511. https://doi.org/10.1016/j.chemosphere.2012.12.012.

18 ISTA (2009). *STA Rules*. Zurich: International Seed Testing Association.

19 Clément, L., Hurel, C., and Marmier, N. (2013). Toxicity of TiO_2 nanoparticles to cladocerans, algae, rotifers and plants – effects of size and crystalline structure. *Chemosphere* 90 (3): 1083–1090. https://doi.org/10.1016/j.chemosphere.2012.09.013.

20 Hashemi, D.E. and Mosavi, M. (2013). Effect of anatase nanoparticles (TiO_2) on parsley seed germination (*Petroselinum crispum*) in vitro. *Biol. Trace Elem. Res.* 155: 283–286.

21 Song, U., Jun, H., Waldman, B. et al. (2013). Functional analyses of nanoparticle toxicity: a comparative study of the effects of TiO_2 and Ag on tomatoes (*Lycopersicon esculentum*). *Ecotoxicol. Environ. Saf.* 93: 60–67. https://doi.org/10.1016/j.ecoenv.2013.03.033.

22 Lee, S., Kim, S., Kim, S., and Lee, I. (2013). Assessment of phytotoxicity of ZnO NPs on a medicinal plant, *Fagopyrum esculentum*. *Environ. Sci. Pollut. Res.* 20 (2): 848–854.

23 Zhao, L., Hernandez-Viezcas, J.A., Peralta-Videa, J.R. et al. (2013). ZnO nanoparticle fate in soil and zinc bioaccumulation in corn plants (*Zea mays*) influenced by alginate. *Environ. Sci. Processes Impacts* 15: 260–266. https://doi.org/10.1039/C2EM30610G.

24 Lee, S., Chung, H., Kim, S., and Lee, I. (2013). The genotoxic effect of ZnO and CuO nanoparticles on early growth of buckwheat, *Fagopyrum esculentum*. *Water Air Soil Pollut.* 224 (9): 1668–1679.

25 Dimkpa, C.O., Latta, D.E., McLean, J.E. et al. (2013). Fate of CuO and ZnO nano- and microparticles in the plant environment. *Environ. Sci. Technol.* 47 (9): 4734–4742. https://doi.org/10.1021/es304736y.

26 Kasana, R.C., Panwar, N.R., Kaul, R.K., and Kumar, P. (2017). Biosynthesis and effects of copper nanoparticles on plants. *Environ. Chem. Lett.* 15: 233–240. https://doi.org/10.1007/s10311-017-0615-5.

27 Adams, J., Wright, M., Wagner, H. et al. (2017). Cu from dissolution of CuO nanoparticles signals changes in root morphology. *Plant Physiol. Biochem.* 110: 108–117. https://doi.org/10.1016/j.plaphy.2016.08.005.

28 Kumar, V., Guleria, P., Kumar, V., and Yadav, S.K. (2013). Gold nanoparticle exposure induces growth and yield enhancement in *Arabidopsis thaliana*. *Sci. Total Environ.* 461: 462–468. https://doi.org/10.1016/j.scitotenv.2013.05.018.

29 Lahiani, M.H., Dervishi, E., Chen, J. et al. (2013). Impact of carbon nanotube exposure to seeds of valuable crops. *ACS Appl. Mater. Interfaces* 5 (16): 7965–7973. https://doi.org/10.1021/am402052x.

30 Serag, M.F., Kaji, N., Habuchi, S. et al. (2013). Nanobiotechnology meets plant cell biology: carbon nanotubes as organelle targeting nanocarriers. *RSC Adv.* 3 (15): 4856–4863.

31 Saifuddin, N., Raziah, A.Z., and Junizah, A.R. (2013). Carbon nanotubes: a review on structure and their interaction with proteins. *J. Chem.* 2013: 676815. https://doi.org/10.1155/2013/676815.

32 Faisal, M., Saquib, Q., Alatar, A.A. et al. (2013). Phytotoxichazards of NiO-nanoparticles in tomato: a study on mechanism of cell death. *J. Hazard. Mater.* 250–251: 318–332. https://doi.org/10.1016/j.jhazmat.2013.01.063.

33 Joseph, T. and Morrison, M. (2008). Nanotechnology in agriculture and food. http://www.nanoforum.org/nf06~module~showmore~folder~99999~scid~377~.html?action=longviewpublication& (accessed 6 June 2008).

34 Feynman, R.P. (1960). There's plenty of room at the bottom. *Eng. Sci.* 23: 22–36. http://www.zyvex.com/nanotech/feynman.html.

35 Hollister, P. (2002). Nanotech: The Tiny Revolution. http://www.nanotech-now.com/CMP-reports/NORWhitePaperJuly2002.pdf (accessed 6 June 2008).

36 Scott, N. and Chen, H. (2012). Industrial Biotechnology. 340–343. https://doi.org/10.1089/ind.2012.1549.

37 Novak, C. and Weatherhead, I. (2004). A microscopic formula for success. *Agriscience News* 2, pp. 1–3. www.syngenta.co.uk/PublicationStore/pdf/Agriscience2.pdf (accessed 9 June 2008).

38 Syngenta (2003). Base-triggered release microcapsules. US Patent 6, 544, 540.

39 Taylor, T.M., Davidson, P.M., Bruce, B.D., and Weiss, J. (2005). Liposomal nanocapsules in food science and agriculture. *Crit. Rev. Food Sci. Nutr.* 45: 587–605.

40 Horiuchi, A., Satou, T., Akao, N. et al. (2005). The effect of free and polyethylene glycol–liposome-entrapped albendazole on larval mobility and number in *Toxocara canis* infected mice. *Vet. Parasitol.* 129: 83–87.

41 Teixeira, M., Alonso, M.J., Pinto, M.M.M., and Barbosa, C.M. (2005). Development and characterization of PLGA nanospheres and nanocapsules containing xanthone and 3-methoxyxanthone. *Eur. J. Pharm. Biopharm.* 59: 491–500.

42 Holister, P., Roman, V.C., and Harper, T. (2008). Nanocapsules. *Technology White Papers* 10. http://www.nanorenac.com/webgestion/intercambio/descargas/Nanocapsules.pdf (accessed 11 June 2008).

43 Harish, P.K.V. and Tharanathan, R.N. (2007). Chitin/chitosan: modifications and their unlimited application potential – an overview. *Trends Food Sci. Technol.* 18: 117–131.

44 Pan, Y., Li, Y.J., Zhao, H.Y. et al. (2002). Bioadhesive polysaccharide in protein delivery system: chitosan nanoparticles improve the intestinal absorption of insulin in vivo. *Int. J. Pharm.* 249: 139–147.

45 Wang, L.Y., Ma, G.H., and Su, Z.G. (2006). Preparation of uniform-sized chitosan microspheres by membrane emulsification technique and application as a carrier of protein drug. *J. Controlled Release* 106: 62–75.

46 Bautista-Banos, S., Hernandez-Lauzardo, A.N., Velázquez-del Valle, M.G. et al. (2006). Chitosan as a potential natural compound to control pre and postharvest diseases of horticultural commodities. *Crop Prot.* 25: 108–118.

47 Trotel-Aziz, P., Couderchet, M., Vernet, G., and Aziz, A. (2006). Chitosan stimulates defense reactions in grapevine leaves and inhibits development of *Botrytis cinerea*. *Eur. J. Plant Pathol.* 114: 405–413.

48 Rabea, E.I., Badawy, M.E.I., Rogge, T.M. et al. (2005). Insecticidal and fungicidal activity of new synthesized chitosan derivatives. *Pest. Manage. Sci.* 61: 951–960.

49 Eweis, M., Elkholy, S.S., and Elsabee, M.Z. (2006). Antifungal efficacy of chitosan and its thiourea derivatives upon the growth of some sugar-beet pathogens. *Int. J. Biol. Macromol.* 38: 1–8.

50 Chien, P.J. and Chou, C.C. (2006). Antifungal activity of chitosan and its application to control post-harvest quality and fungal rotting of Tankan citrus fruit (*Citrus tankan Hayata*). *J. Sci. Food Agric.* 86: 1964–1969.

51 Goldwasser, Y., Eizenberg, H., Golan, S., and Kleifeld, Y. (2003). Control of *Orobanche crenata* and *Orobanche aegyptiaca* in parsley. *Crop Prot.* 22: 295–305.

52 Joel, D.M. (2000). The long-term approach to parasitic weeds control: manipulation of specific developmental mechanisms of the parasite. *Crop Prot.* 19: 753–758.

53 Joel, D.M., Hershenhorn, J., Eizenberg, H. et al. (2007). Biology and management of weedy root parasites. *Hortic. Rev.* 33: 267–349.

54 Jurado-Exposito, M., García-Torres, L., and Castejon-Muñoz, M. (1997). Broad bean and lentil seed treatments with imidazolines for the control of broomrape (*Orobanche crenata*). *J. Agric. Sci.* 129: 307–314.

55 Wiesman, Z., Ben, D.N., Sharvit, E. et al. (2007). Novel cationic vesicle platform derived from vernonia oil for efficient delivery of DNA through plant cuticle membranes. *J. Biotechnol.* 130: 85–94.

56 López-Ráez, J.A., Matusova, R., Cardoso, C. et al. (2009). Strigolactones: ecological significance and use as a target for parasitic plant control. *Pest. Manage. Sci.* 64: 471–477.

57 Vurro, M., Boari, A., Evidente, A. et al. (2009). Natural metabolites for parasitic weed management. *Pest. Manage. Sci.* 65: 566–571.

58 Holister, P., Weener, J.W., Roman, V.C., and Harper, T. (2008). Nanoparticles. *Technology White Papers* 3. http://www.nanorenac.com/webgestion/intercambio/descargas/Nanoparticles.pdf (accessed 12 June 2008).

59 Grover, A., Mittal, D., and Negi, M. (2013). Generating high temperature tolerant transgenic plants: achievements and challenges. *Plant Sci.* 205–206: 38–47.

60 Grover, A., Kapoor, A., and Lakshmi, O.S. (2001). Understanding molecular alphabets of the plant abiotic stress responses. *Curr. Sci.* 80: 206–216.

61 Lang, S.S. (2013). Waste fiber can be recycled into valuable products using new technique of electrospinning. Cornell Researchers Report. http://www.news.cornell.edu/stories/2003/09/electrospinning-cellulose-waste-fiber (accessed 5 February 2021).

62 Awizar, D.A., Othman, N.K., Jalar, A. et al. (2013). Nanosilicate extraction from rice husk ash as green corrosion inhibitor. *Int. J. Electrochem. Sci.* 8: 1759–1769.

63 Sekhon, B.S. (2014). Nanotechnology in agri-food production: an overview. *Nanotechnol. Sci. Appl.* 7: 31–53. https://doi.org/10.2147/NSA.S39406.

64 Nair, R., Varghese, S.H., and Nair, B.G. (2010). Nanoparticulate material delivery to plants. *Plant Sci.* 179: 154–163. https://doi.org/10.1016/j.plantsci.2010.04.012.

65 Torney, F., Trewyn, B.G., and Lin, V.S.Y. (2007). Mesoporous silica nanoparticles deliver DNA and chemicals into plants. *Nat. Nanotechnol.* 2: 295–300. https://doi.org/10.1038/nnano.2007.108.

66 Das, R.K. and Brar, S.K. (2013). Plant mediated green synthesis: modified approaches. *Nanoscale* 5 (21): 10155–10162.

67 Narayanan, K.B. and Sakthivel, N. (2010). Biological synthesis of metal nanoparticles by microbes. *Adv. Colloid Interface Sci.* 156 (1): 1–13.

68 Thakkar, K.N., Mhatre, S.S., and Parikh, R.Y. (2010). Biological synthesis of metallic nanoparticles. *Nanomed. Nanotechnol. Biol. Med.* 6 (2): 257–262.

69 Mandal, D. (2006). The use of microorganisms for the formation of metal nanoparticles and their application. *Appl. Microbiol. Biotechnol.* 69 (5): 485–492.

70 Gericke, M. and Pinches, A. (2006). Biological synthesis of metal nanoparticles. *Hydrometallurgy* 83 (1): 132–140.

71 Mohanpuria, P., Rana, N.K., and Yadav, S.K. (2008). Biosynthesis of nanoparticles: technological concepts and future applications. *J. Nanopart. Res.* 10 (3): 507–517.

72 Hulkoti, N.I. and Taranath, T. (2014). Biosynthesis of nanoparticles using microbes – a review. *Colloids Surf., B* 121: 474–483.

73 Das, R.K., Pachapur, V., Lonappan, L. et al. (2017). Biological synthesis of metallic nanoparticles: plants, animals and microbial aspects. *Nanotechnol. Environ. Eng.* 2 https://doi.org/10.1007/s41204-017-0029-4.

74 Mann, S., Frankel, R.B., and Blakemore, R.P. (1984). Structure, morphology and crystal growth of bacterial magnetite. *Nature* 310 (5976): 405.

75 Smith, P.R. (1998). Photophysical and photochemical characterization of bacterial semiconductor cadmium sulfide particles. *J. Chem. Soc., Faraday Trans.* 94 (9): 1235–1241.

76 Joerger, R., Klaus, T., and Granqvist, C. (2000). Biologically produced silver-carbon composite materials for optically functional thin-film coatings. *Adv. Mater.* 12 (6): 407–409 144.

77 Nair, B. and Pradeep, T. (2002). Coalescence of nanoclusters and formation of submicron crystallites assisted by *Lactobacillus* strains. *Cryst. Growth Des.* 2 (4): 293–298.

78 Zhang, H. (2005). Biosorption and bioreduction of diamine silver complex by *Corynebacterium*. *J. Chem. Technol. Biotechnol.* 80 (3): 285–290.

79 Lengke, M.F., Fleet, M.E., and Southam, G. (2006). Synthesis of platinum nanoparticles by reaction of filamentous cyanobacteria with platinum (IV)-chloride complex. *Langmuir* 22 (17): 7318–7323.

80 Prasad, K., Jha, A.K., and Kulkarni, A. (2007). Lactobacillus assisted synthesis of titanium nanoparticles. *Nanoscale Res. Lett.* 2 (5): 248–250.

81 Kalimuthu, K., Suresh, R.B., Venkataraman, D. et al. (2008). Biosynthesis of silver nanocrystals by *Bacillus licheniformis*. *Colloids Surf., B* 65 (1): 150–153.

82 Babu, M.M.G. and Gunasekaran, P. (2009). Production and structural characterization of crystalline silver nanoparticles from *Bacillus cereus* isolate. *Colloids Surf., B* 74 (1): 191–195.

83 Kim, S.W., Kim, K.S., Lamsal, K. et al. (2009). An in vitro study of the antifungal effect of silver nanoparticles on oak wilt pathogen *Raffaelea* sp. *J. Microbiol. Biotechnol.* 19: 760–764.

84 Kasprowicz, M.J., Kozioł, M., and Gorczyca, A. (2010). The effect of silver nanoparticles on phytopathogenic spores of *Fusarium culmorum*. *Can. J. Microbiol.* 56: 247–253.

85 Du, L., Jiang, H., Liu, X., and Wang, E. (2007). Biosynthesis of gold nanoparticles assisted by *Escherichia coli* DH5α and its application on direct electrochemistry of hemoglobin. *Electrochem. Commun.* 9 (5): 1165–1170.

86 Bao, H., Lu, Z., Cui, X. et al. (2010). Extracellular microbial synthesis of biocompatible CdTe quantum dots. *Acta Biomater.* 6 (9): 3534–3541.

87 Mukherjee, P. (2002). Extracellular synthesis of gold nanoparticles by the fungus *Fusarium oxysporum*. *ChemBioChem* 3 (5): 461–463.

88 Bansal, V. (2004). Biosynthesis of zirconia nanoparticles using the fungus *Fusarium oxysporum*. *J. Mater. Chem.* 14 (22): 3303–3305.

89 Bansal, V. (2006). Room-temperature biosynthesis of ferroelectric barium titanate nanoparticles. *J. Am. Chem. Soc.* 128 (36): 11958–11963.

90 Vigneshwaran, N. (2007). Biological synthesis of silver nanoparticles using the fungus *Aspergillus flavus*. *Mater. Lett.* 61 (6): 1413–1418.

91 Sanghi, R. and Verma, P. (2009). Biomimetic synthesis and characterization of protein capped silver nanoparticles. *Bioresour. Technol.* 100 (1): 501–504.

92 Senapati, S. (2005). Extracellular biosynthesis of bimetallic Au–Ag alloy nanoparticles. *Small* 1 (5): 517–520.

93 Riddin, T., Gericke, M., and Whiteley, C. (2006). Analysis of the interand extracellular formation of platinum nanoparticles by *Fusarium oxysporum* f. sp. *lycopersici* using response surface methodology. *Nanotechnology* 17 (14): 3482.

94 Balakumaran, M.D., Ramachandran, R., and Kalaicheilvan, P.T. (2015). Exploitation of endophytic fungus, *Guignardia mangiferae* for extracellular synthesis of silver nanoparticles and their in vitro biological activities. *Microbiol. Res.* 178: 9–17. https://doi.org/10.1016/j.micres.2015.05.009.

95 Rose, G.K., Soni, R., Rishi, P., and Soni, S.K. (2019). Optimization of the biological synthesis of silver nanoparticles using *Penicillium oxalicum* GRS-1 and their antimicrobial effects against common food-borne pathogens. *Green Process. Synth.* 8: 144–156. https://doi.org/10.1515/GPS-2018-0042.

96 Costa Silva, L.P., Oliveira, J.P., Keijok, W.J. et al. (2017). Extracellular biosynthesis of silver nanoparticles using the cell-free filtrate of nematophagous fungus *Duddingtonia flagans*. *Int. J. Nanomed.* 12: 6373–6381. https://doi.org/10.2147/IJN.S137703.

97 Ahluwalia, V., Kumar, J., Sisodia, R. et al. (2014). Green synthesis of silver nanoparticles by *Trichoderma harzianum* and their bio-efficacy evaluation against *Staphylococcus aureus* and *Klebsiella pneumoniae*. *Ind. Crops Prod.* 55: 202–206. https://doi.org/10.1016/j.indcrop.2014.01.026.

98 Kashyap, P.L., Kumar, S., Srivastava, A.K., and Sharma, A.K. (2013). Myconanotechnology in agriculture: a perspective. *World J. Microbiol. Biotechnol.* 29: 191–207. https://doi.org/10.1007/s11274-012-1171-6.

99 Luque, A.P. and Rubiales, D. (2009). Nanotechnology for parasitic plant control. *Pest. Manage. Sci. J.* 65: 540–545.

100 Li, Z.Z., Chen, J.F., Liu, F. et al. (2007). Study of UV-shielding properties of novel porous hollow silica nanoparticle carriers for avermectin. *Pest. Manage. Sci.* 63: 241–246.

101 Roe, D., Karandikar, B., Bonn-Savage, N. et al. (2008). Antimicrobial surface functionalization of plastic catheters by silver nanoparticles. *J. Antimicrob. Chemother.* 61: 869–876.

102 Oh, S.D., Lee, S., Choi, S.H. et al. (2006). Synthesis of Ag and Ag-SiO_2 nanoparticles by γ-irradiation and their antibacterial and antifungal efficiency against *Salmonella enteric* serovar Typhimurium and *Botrytis cinerea*. *Colloids Surf., A* 275: 228–233.

103 Gajbhiye, M., Kesharwani, J., Ingle, A. et al. (2009). Fungus-mediated synthesis of silver nanoparticles and their activity against pathogenic fungi in combination with fluconazole. *Nanomedicine* 5: 382–386.

104 Jo, Y.K., Kim, B.H., and Jung, G. (2009). Antifungal activity of silver ions and nanoparticles on phytopathogenic fungi. *Plant Dis.* 93: 1037–1043.

105 Min, J.S., Kim, K.S., Kim, S.W. et al. (2009). Effects of colloidal silver nanoparticles on sclerotium-forming phytopathogenic fungi. *Plant Pathol. J.* 25: 376–380.

106 Musarrat, J., Dwivedi, S., Singh, B.R. et al. (2010). Production of antimicrobial silver nanoparticles in water extracts of the fungus *Amylomyces rouxii* strain KSU-09. *Bioresour. Technol.* 101: 8772–8776.

107 Roy Choudhury, S., Ghosh, M., Mandal, A. et al. (2011). Surface-modified sulfur nanoparticles: an effective antifungal agent against *Aspergillus niger* and *Fusarium oxysporum*. *Appl. Microbiol. Biotechnol.* 90: 733–743. https://doi.org/10.1007/s00253-011-3142-5.

108 Choudhury, S.R., Nair, K.K., Kumar, R. et al. (2011). Surface-modified sulfur nanoparticles: an effective antifungal agent against *Aspergillus niger* and *Fusarium oxysporum*. *Appl. Microbiol. Biotechnol.* 90: 733–743.

109 González-Melendi, P., Fernández-Pacheco, R., Coronado, M.J. et al. (2008). Nanoparticles as smart treatment-delivery systems in plants: assessment of different techniques of microscopy for their visualization in plant tissues. *Ann. Bot.* 101: 187–195.

110 Khan, M.R. and Anwer, A. (2011). Fungal bio inoculants for plant disease management. In: *Microbes and Microbial Technology* (eds. M. Paul, M. Clinton and I. Ahmad), 447–448. New York: Springer.

111 Khan, M.R. and Haque, Z. (2013). Morphological and biochemical responses of five tobacco cultivars to simultaneous infection with *Pythium aphanidermatum* and *Meloidogyne incognita*. *Phytopathol. Mediterr.* 52: 98–109.

112 Wang, Z., Wei, F., Liu, S.Y. et al. (2010). Electrocatalytic oxidation of phytohormone salicylic acid at copper nanoparticles-modified gold electrode and its detection in oilseed rape infected with fungal pathogen *Sclerotinia sclerotiorum*. *Talanta* 80: 1277–1281.

113 Gogoi, R., Dureja, P., and Singh, P.K. (2009). Nanoformulations a safer and effective option for agrochemicals. *Indian Farming* 59: 7–12.

114 Mahendra, R., Shivaji, D., Aniket, G., and Elsalam, K.A. (2012). Strategic nanoparticle-mediated gene transfer in plants and animals – a novel approach. *Curr. Nanosci.* 8: 170–179.

115 Biswal, S.K., Nayak, A.K., Parida, U.K., and Nayak, P.L. (2012). Applications of nanotechnology in agriculture and food sciences. *Int. J. Sci. Innovation Discovery* 2: 21–36.

116 Sharon, M., Choudhary, A.K., and Kumar, R. (2010). Nanotechnology in agricultural diseases and food safety. *J. Phytol.* 2: 83–92.

4

Nanotechnology in Agri-Food Production

CHAPTER MENU

4.1 Nanomaterials

Nanotechnology has great potential to generate materials with specific properties and develop innovative products [1, 2]. It involves nanoscale processing, handling, use, and characterization of materials. Any material of one or more external dimensions with 1–100 nm [3] is classified as such. The idea of a nanomaterial can depend on the organization, agency, or committee for various purposes [4]. The concept can only be defined on the basis of the number of dimensions, not just for nanoscale ranges. It can find measurements of their structural elements on this definition: (i)zero dimensional (0D) – where all nanoscale sizes include nanocluster materials, nanoparticles, and nanodispersions; (ii) one dimensional (1D) – nanofibers and nanotubular materials with rod or tube lengths ranging from 100 nm to a few micrometres and where one dimension is beyond the nanoscale; (iii) two-dimensional (2D) – a plate-like structure that can be nanometre-thick films and coatings and where two of the measurements are not limited to the nanoscale; and (iv) three dimensional (3D) – materials where the structural elements of 0D, 1D, and 2D can be present in close contact with each other and form interfaces [5–7].

Nano-Technological Intervention in Agricultural Productivity, First Edition.
Javid A. Parray, Mohammad Yaseen Mir and Nowsheen Shameem.

4.2 Nanotechnology and Food Systems: Food Packing

The interest in nanomaterials' use is linked to their unique characteristics, which can only be done because of their nanoscale and vary significantly from their micro- and macroscopic counterparts, indicating significant advantages for packaging materials in many situations. The nanoscale effect on materials and systems is often linked to their large surface-to-volume ratio, resulting in increased surface area, differentiated optical behaviour, chemical and kinetic stability, and low density versus high mechanical properties [6].

These micro- and macroscale advantages lead to great interest in their use by the packaging industry. They can offer many benefits, such as improved mechanical and barrier properties, thermal stability, and improved optical properties. Nanotechnology in food packaging has shown great potential to enhance packaging materials' critical properties and introduce new features, such as active and intelligent features. It is also possible to use nanotechnology to solve some of the key issues with traditional packaging. In food packaging, the necessary regulations for materials and articles intended to come in contact with food are laid down by the European Union (EU) and requirements are provided for engineered nanomaterials [8, 9].

The primary materials used in food packaging are polymers. It is possible to enhance these materials' properties by adding other compounds, such as nanomaterials, into the polymer matrix. The silicate presence in a polymer polyamide matrix contributes to creating a high mechanical strength nanocomposite material with excellent high-temperature characteristics [10, 11]. The possibility of the nanoscale use of silver in resin to develop packaging materials with antibacterial properties was stated in 1997 [11]. Subsequently, the thermoplastic moulding material containing polyamide, nanosized fillers, and elastomers used in the preparation of mono- or multilayer films or hollow bodies was referred to for the packaging foodstuffs, such as meat sausages, cheese, and beverages [12]. Silver nanoparticles are presented in many plastic food containers, and packaging materials are sold by companies such as Kinetic, Always New, and Fresher Longer. The use of antimicrobial silver nanoparticles enables food shelf life to be extended [13].

Their mechanical effect facilitates the integration of nanomaterials into packaging materials, i.e. high strength and rigidity in addition to low permeability properties. Because of their ability to enhance the barrier, thermal, and mechanical properties, both clay and silver nanoparticles are the most used nanomaterials in packaging. However, ZnO and TiO nanoparticles have also been shown to be desirable alternatives for improving barrier ability and obtaining antimicrobial properties because of their potential antimicrobial activity. The use of bio-based and biodegradable nanomaterials in nanocomposites for packaging applications is also the subject of attention [14]. They have different characteristics of inorganic and metallic materials. Still, they can open the door to many possibilities in this field because of their non-toxicity, biodegradability, and receptive behaviour. The starch and chitosan nanostructures have been successfully used to improve packaging materials [15, 16]. To improve packaging materials, both nano-fibrillated cellulose (NFC) and cellulose nanocrystals (CNCs) were used to obtain packaging materials with multifunctional capabilities [17, 18].

The packaging is often used to connect with customers and across the food chain and ensure traceability, including ingredients and nutritional information. As a medium for product marketing, packaging design is critical in consumers' decisions [19]. In this way, nanotechnology will help the industry produce advanced packaging materials, enable the manufacture of new feature packaging systems, and nanomaterials allow the manufacture of nanocomposites, leading to improved packaging materials' properties. One of the real areas of interest is nanostructures' use for the encapsulation of active compounds and their inclusion in packaging materials, envisaging the production of dynamic and smart packaging materials. Encapsulation would facilitate the defence of active compounds such as antifungal and antibacterial agents, antioxidants, and oxygen scavengers, thus helping to preserve their properties for more extended periods, while at the same time enabling the properties of the material used for the development of nanostructures to be combined with those of the packaging material itself. This would make it possible to use biologically active molecules such as essential oils, peptides, and other molecules that cannot commonly be added during packaging [20–23].

In the production of materials with improved physical, chemical, and biological (antimicrobial) properties, graphene has been listed as one of the materials of interest for developing packaging materials [24–26]. Graphene can also produce sensors that can be used to create smart packaging through food spoilage detection indicators to warn the customer about food quality [27]. Intelligent packaging also offers considerable potential as a marketing tool and establishes brand differentiation, apart from aspects of consistency, protection, and distribution [28, 29]. Applications of this type using nanotechnology have just begun to be explored, and a smart sticker for use with perishable foods [30, 31] and an oxygen detection device [32, 33] have recently been presented along with printed electronics for sensor development [34].

4.3 Nano-Nutraceuticals

Developing promising nutraceuticals in the human body with stronger biokinetic and biodynamic reactions is doing wonders now. Nutraceutical products are typically administered by oral intake, unlike medications, leading to low absorption through nutraceutical products' gastrointestinal tract. Some products are available, such as milk thistle, which has liver detoxification effects, and much of its active ingredient, silymarin, is lost after ingestion. It is also not bioavailable [35]. Many nutraceuticals containing antioxidant elements, such as α-tocopherol and ascorbic acid, are also typically unstable, poorly soluble in water, and poorly distributed in vivo. There is also a need to deliver nutraceutical products by also making them more bioavailable more effectively. More enthusiasm has been generated by the possible contributions of nanoscale phenomena to enhance human health, leading to new biomaterials, devices, and biological detection and remediation techniques. A recent application of nutritional medicine to improve nutrient absorption and solubility is nanotechnology in enhancing nutraceuticals' delivery. By reducing size, the derived properties can help modify the complex biological process into a desirable approach. For instance, silymarin has been reformulated into nano-liposomes; significant absorption increases have been reported [36].

Nano-liposomes made from phospholipids have attracted great interest among the nanoparticles developed using phospholipids, lactic acid, chitosan, dextran, polyethylene glycol, cholesterol, carbon, silica, and some metals as efficient carriers for nutrients, drugs, and other bioactive agents. Mozafari and Mortazavi [37] identified liposomes as a bilayer of phospholipid-based liquid-filled bubbles. Interactions between the aqueous medium and amphiphilic lipids are the basis of the mechanism of liposome formation. The aqueous stage of both the internal and external media is exposed to the phospholipid polar head groups, and the hydrophobic hydrocarbon tails are connected with the bilayer because of the specific phospholipid composition that happens. The two-layer liposomes' arrangement is similar to that of the membranes of the human body cells surrounding each cell.

Lipophilic molecules (e.g. cyclosporine A, vitamin E, and phenytoin) can also be integrated into liposomal membranes or may be compounded with cyclodextrins and then encapsulated in an aqueous liposome compartment [38]. Suntres and Shek [31] demonstrated that paraquat-induced lung injury is alleviated by pretreatment with α-tocopherol and glutathione-containing bifunctional liposomes. The targeted delivery of both in vivo and in vitro conditions via active and passive mechanisms [37] is another property of lipid-based nanocarriers.

At present, the most potent oral way to distribute nutrients is via liposomes. The stomach flora must be digested by orally administered nutrients and making it pass through the liver's first passage before entering the bloodstream. Another characteristic of a liposome is that it not only gets its material into serum blood but can also get it into the cells where it is required directly. There are some commercially available supplements for nutrition that use delivery mechanisms based on liposomes. In fortified orange juice, Marsanasco et al. [33] discussed liposomes as vehicles for vitamins E and C and also discussed vitamin C defence after heat treatment. The liposomes based on soy phosphatidylcholine were tested as vitamin transporters to be introduced into orange juice and then subjected to pasteurization. The combination of liposomal formulations and orange juice vitamins did not change its organoleptic properties and, after pasteurization and storage at 4 °C for 37 days, demonstrated microbiological stability. Gonnet et al. [39] stated that deliberate encapsulation of liposoluble vitamins (A, D, E, and K) and carotenoids leads to increased efficacy, allowing lower doses to be administered, minimizing the possible symptoms and side effects of hypervitaminosis syndrome and offering many health benefits. In situations where fat-soluble vitamin absorption is impaired, Kirilenko and Gregoriadis [40] reveal the possible role of liposomes in promoting fat-soluble vitamins after oral administration.

4.3.1 Issues with Nano-Nutraceuticals

Although there is no doubt that nanotechnology applications are potentially useful and far-reaching, the protection of this technology has not been developed. The Food and Drug Administration (FDA) has created a nanotechnology task force to resolve safety and regulatory issues in response to nanotechnology's growing use in foods, supplements, and cosmetics. In 2012, the FDA released draught guidelines on nanotechnology [41], focusing on food and cosmetics. Although neither document directly discusses dietary supplements, they are considered by the FDA to be a form of food. Draught food guidance is likely to have a significant effect on the dietary supplement industry. However, FDA cites nanotechnology as an example of a method that creates a new nutrient in the recently released New Dietary Ingredient (NDI) draught guidelines and may require FDA notification, resulting in new or altered chemical properties of the compound. There is no doubt that nanotechnology innovations offer unprecedented advantages in the fields of personal care and nutrition. However, there are still so many unanswered questions concerning the security of this new technology. The most ardent exponents of nanotechnology's advancement accept that the widespread use of nanomaterials poses unknown risks.

4.4 Nanotechnological Advancement in Antimicrobial Peptides (AMPs)

Nanotechnology uses two primary approaches to encapsulate antimicrobial peptides (AMPs). Passive delivery is known as undirected delivery. It involves conventional nano-delivery systems that do not have any structural modification to the surface, which can be managed by manipulating the nanocarrier's size and shape [42]. Direct distribution is referred to as robust targeting. The nanocarrier's surface chemistry with the ligands or other moieties is changed to permit interaction with the intended site [43]. Both delivery systems have many benefits and inconveniences.

Passive systems tend to have fewer agents in their composition than an active system. Often, a passive system encapsulates peptides without any further surface modification. On the other hand, a dynamic system contains modified ligands or other surface transport drives to facilitate their association with infected cells and increase the transport of drugs in certain unique regions [44, 45].

4.4.1 Passive Nano-Delivery Systems

The AMPs that are being tested whether in vitro, in vivo, or both, as potential antimicrobials, and a few clinical studies for such peptides have been published. For different peptides, as mentioned below, the passive delivery mechanism is discussed.

4.4.1.1 Cyclosporin A

A non-ribosomal, cyclic deca-peptide composed of hydrophobic amino acid residues is isolated from the fungus *Tolypocladium inflatum*. It has been used to prevent the nephrotic syndrome, Crohn's refractory disorder, ulcerative colitis, aplastic anaemia, myasthenia gravis, dermatomyositis, biliary cirrhosis, psoriasis, rheumatoid arthritis, etc. [46–48]. The peptide is lipophilic and, to improve its bioavailability and reduce its collateral effects, can be easily incorporated into nano-delivery systems [46, 49, 50]. Cyclosporin A was evaluated for enhanced dermal bioavailability in the ex vivo setting in amorphous nanoparticles and was significantly increased its penetration by approximately six times over the free peptide [50]. Other nanoparticles charged with chitosan peptides showed improved pharmacokinetics to boost cyclosporin A's bioavailability (1.8-fold increase) in the time concentration of commercial Neoral-R microemulsion [51].

4.4.1.2 Nisin

Nisin is an enzyme-mediated post-translational modification bacteriocin containing rare amino acids [52]. It belongs to the lantibiotic class and includes dry serine and threonine residues resulting in dehydroalanine and dehydrobutyrine [53–55]. Nisin has a wide range of Gram-positive bacteria and is usually non-cytotoxic to mammals and helps prolong food shelf life. The nisin was loaded into a nano-delivery device to enhance their bioavailability for its role in milk fermentation; it has been encapsulated in liposomes [56]. Thus, these nanocarriers provide an alternative food preservative for cheese production with safe and prolonged release of peptides that could give extended shelf life and quality.

The nisin peptide was encapsulated into polymeric polycaprolactone nanoparticles [57]. Similar antimicrobial activity against *Candida albicans* strains was demonstrated by encapsulated nisin as free nisin, suggesting the ability of encapsulated nisin as a prophylactic against candidiasis and other gynaecological infections [57].

4.4.1.3 Polymyxin

Polymyxin is a peptide isolated from Bacillus poly mix that inhibits a wide range of Gram-negative bacteria. Its mechanism is attachment to lipopolysaccharides of the acidic phospholipids and bacterial cell membrane, leading to the leakage of intracellular components resulting in death [58]. For the treatment of urinary tract, meningitis, mucositis, otitis, periodontitis, lung, skin, eye, and lesions, polymyxin B has been extensively used [59, 60].

To overcome bioavailability and decrease the neighbouring effects of polymyxin, it was also nano-formulated. They investigated polymyxin's antimicrobial role through its in vitro encapsulation of solid lipid nanoparticles (SLPs). This peptide demonstrated its activity at only 10 mg/ml, while nanoparticles loaded with polymyxin were active at a much lower concentration, introducing them as a novel antibiotic drug therapy [61]. These AMPs are encapsulated and tested in the liposome against resistant bacterial strains. Polymyxin-loaded liposomes display high cell absorption by immunocytochemistry studies [62]. Besides, immunocytochemistry and microscopy studies have shown that resistant *Pseudomonas aeruginosa* has a high penetration rate of encapsulated polymyxin B [62] compared to its free form. Therefore, to encapsulate polymyxin, nanocarriers have been applied, presenting promising results.

4.4.2 Antimicrobial Peptides in Targeted Nano-Delivery Systems

To enhance their delivery to the intended sites by more precise delivery of peptides of interest, nanosystems were decorated or updated with other moieties [44, 63]. This mechanism helps resolve natural obstacles, including low pH and enzymes in the gastrointestinal tract and the mucosa and epithelial layers of the intestine [64].

In a targeted delivery method utilizing polymeric nanoparticles, Sneh-Edri et al. [65] experimented with a synthetic peptide. Using two lines of mouse dendritic cells, namely, DC 2.4 and B3Z hybridoma T cells, capable of recognizing the peptide sequence of interest [65], their research investigated the absorption and intracellular targeted drug delivery. Increased absorption was explicitly documented based on peptide binding to T cell receptors, enabling the marker's delivery into the cells, an exciting site-specific treatment strategy. In another study, a modified fusogenic liposome was used to carry vancomycin. Because of its inability to cross the bacterial cell membranes, the peptide is not recognized as active against Gram-negative bacteria. [66]. Vancomycin was encapsulated in phospholipid cholesterol hemisuccinate-covered tiny unilamellar liposomes with an affinity for the eukaryote cell membrane. The antimicrobial activity against clinical isolates [66] was then tested. The liposomes' surface modification enhanced the efficiency of vancomycin against the pathogens.

Additionally, to allow the delivery of the drug of interest, AMPs may be adsorbed on the surface of the nanocarrier. For example, CDX, which was conjugated with a neurotoxin-derived peptide, was also reported for delivery to the brain of an anticancer agent, doxorubicin, in red blood cell membrane-coated poly(lactic-co-glycolic acid) (PLGA) nanoparticles. CDX can do in-derived peptide found on the brain's endothelial surface with a high binding affinity to nicotinic acetylcholine receptors. A red blood cell membrane was coated with this peptide, which was also painted on a PLGA nanoparticle surface to facilitate the blood–brain barrier passage. When providing doxorubicin in vitro, nanoparticles showed promising efficiency with primary brain capillary cells isolated from the Wistar rats and U87 human brain cells. In addition to destroying microorganisms, AMPs have been screened for cancerous tumours [67, 68].

4.5 Assessment of Nanotechnology for Enhanced Food Security

Rising R&D costs, the lack of qualified human capital, and the policy structure for incorporating emerging technologies into applications across the agri-value chain are significant challenges for policymakers and technology developers worldwide. The current emphasis is on exploring and understanding the potential role of recent technological advances in nanotechnology in improving food and nutrition available in the post-harvest and food processing phases [69]. Through the query-based system, the definition of technology road mapping and database management combines nanoscience to map the potential of such technologies against the existing information gaps in nanotechnology to enhance food quality and quantity.

4.5.1 Framework for Assessing the Potential Role of Nanotechnology in Food

The developed framework for Incorporating Research in Nanotechnology in the Food Sector aims to map nano-research areas into the food supply chain, i.e. the entire production consumption system, first by defining the nano-research scope and the food supply chain and then by connecting nano-research to the food supply chain. Different nano-research fields and food supply chain sectors are defined using structured search strings and a technology-based process methodology [69].

4.5.2 Assessment of Nanotechnology Potential Through Literature Survey

A thorough review of the data will help determine and classify essential testing tools [70] that can be incorporated into the food application in each of these regions. Efforts have therefore been made to define and map the areas of nano-research. The food supply chain to the food sector determinants mentioned earlier, namely, food processing, packaging, nutraceutical distribution, protection and sensing, and functional foods, and some nanotechnology areas have been defined based on the information mapping system with potential applications in food sector. These involved the processing of food, packaging, quality control, and delivery of nutraceuticals and functional foods. Nano-research areas ranged from formulations such as gels, emulsions to particles, and devices and were diverse. Applications of several innovations were detailed and included several fields. Most of them were gradual and complementary gaps in the current level of expertise to enhance the goods' distribution, shelf life, or product quality.

Approximately 6% of the published literature in collecting collected documents was discovered by research organizations based in India, suggesting that the country has also initiated the current focus on nanotechnology's potential application to this field, which can deduce that early signs of India's expenditure in various grants and schemes have begun to give rise to early-stage innovations. New trends must be catalysed by constructive measures to change the R&D environment, implemented by policy action, so that investments are designed prudently, including the development of skilled human resources. India's investments in nanotechnology are also smaller than at the global level. Roughly, US$20 billion was spent in 2008 from both the public and private sectors worldwide. Market estimates show an investment of US$1880 billion in nanotechnology-related revenues across all industries by 2015.

4.6 Future Perspectives

Before the impact of exposure to nanoparticles on human health is known, several factors must be considered. A precautionary approach and research in the field of nanotoxicology [71] are essential. Besides, in the Indian context, there is a need to involve stakeholders in an agri-value chain from the early stages of technology growth to be aware of the associated risks and uncertainties [70]. These interventions will contribute to better understanding and acceptance of research products and catalyse technology production's life cycle. There will be more need for a coordinated risk management approach as more innovations flow from this emerging field. The evolution of the Indian agriculture and food system's participatory, dynamic, and sensitive nanotechnology policy and procedure will be necessary if this enabling technology's economic impact is to hit the agricultural society.

References

1 Parisi, C., Vigani, M., and Rodríguez-Cerezo, E. (2015). Agricultural nanotechnologies: what are the current possibilities? *Nano Today* 10: 124–127.

2 Neethirajan, S. and Jayas, D.S. (2011). Nanotechnology for the food and bioprocessing industries. *Food Bioprocess Technol.* 4: 39–47.

3 Potocnik, J. (2011). Commission recommendation of October 18 2011 on the definition of nanomaterial (2011/696/EU). *Off. J. Eur. Union* L275: 38–40.

4 Rauscher, H., Robben, G., Amenta, V. et al. (2014). Towards a Review of the EC Recommendation for a Definition of the Term "Nanomaterial" Part 1: Compilation of Information Concerning the Experience with the Definition. Publications Office of the European Union. Doi https://doi.org/10.2788/36343.

5 Sanjay, S.S. and Pandey, A.C. (2017). A brief manifestation of nanotechnology. In: *EMR/ESR/EPR Spectroscopy for Characterization of Nanomaterials*, Advanced Structured Materials, vol. 62 (ed. A.K. Shukla), 47–63. Springer.

6 Gleiter, H. (2000). Nanostructured materials: basic concepts and microstructure. *Acta Mater.* 48: 1–29.

7 Tiwari, J.N., Tiwari, R.N., and Kim, K.S. (2012). Zero-dimensional, one-dimensional, two-dimensional and three-dimensional nanostructured materials for advanced electrochemical energy devices. *Prog. Mater. Sci.* 57: 724–803.

8 European Commission. Commission regulation (EU) no 10/2011 on plastic materials and articles intended to come into contact with food. *Off. J. Eur. Union* 15: 12–88.

9 Grand View Research (2016). Nano-enabled Packaging Market Report Nano-enabled Packaging Market Analysis By Technology (Active Packaging, Intelligent & Smart Packaging, Controlled Release Packaging), By Application (Food & Beverages, Pharmaceutical, Personal Care & Cosmetics) And Segment Forecast to 2024, pp. 120. http://www.grandviewresearch.com/industry-analysis/nano-enabled-packaging-market (accessed 01 February 2021).

10 Okada, A., Fukushima, Y., Kawasumi, M. et al. (1988). Composite material and process for manufacturing same. US patent 4739007. https://www.google.com/patents/US4739007 (accessed 01 February 2021).

11 Maxfield, M. and Christiani, B.R. (1993). Polymer nanocomposites formed by melt processing of a polymer and an exfoliated layered material derivatized with reactive organo silanes. Patent 4,739,007.

12 Ulrich, R., Mahlke, M., Schaefer, M. et al. (2004). Thermoplastic molding material containing polyamide, nanoscale filler and elastomer useful in the preparation of mono- or multilayer films or hollow bodies, for packaging foodstuffs, such as meat sausages, cheese, drinks. DE2002137820 20020819.

13 Nanocor, IMPERM. (2020). Advanced Performance Additives Specialty Materials Nanocor. https://www.mineralstech.com/business-segments/performance-materials/advanced-performance-additives.

14 Rhim, J.W.W., Park, H.M.M., and Ha, C.S.S. (2013). Bio-nanocomposites for food packaging applications. *Prog. Polym. Sci.* 38: 1629–1652.

15 de Moura, M.R., Aouada, F.A., Avena-Bustillos, R.J. et al. (2009). Improved barrier and mechanical properties of novel hydroxypropyl methylcellulose edible films with chitosan/tripolyphosphate nanoparticles. *J. Food Eng.* 92: 448–453.

16 Le Corre, D., Bras, J., and Dufresne, A. (2010). Starch nanoparticles: a review. *Biomacromolecules* 11: 1139–1153.

17 Abdollahi, M., Alboofetileh, M., Rezaei, M., and Behrooz, R. (2013). Comparing physicomechanical and thermal properties of alginate nanocomposite films reinforced with organic and/or inorganic nanofillers. *Food Hydrocolloids* 32: 416–424.

18 Bilbao-Sainz, C., Bras, J., Williams, T. et al. (2011). HPMC reinforced with different cellulose nanoparticles. *Carbohydr. Polym.* 86: 1549–1557.

19 Simmonds, G. and Spence, C. (2016). Thinking inside the box: how seeing products on, or through, the packaging influences consumer perceptions and purchase behaviour. *Food Qual. Preference* https://doi.org/10.1016/j.foodqual.2016.11.010.

20 Yam, K.L. and Lee, D.S. (2012). Emerging food packaging technologies: an overview. In: *Emerging Food Packaging Technologies: Principles and Practice* (eds. K.L. Yam and D.S. Lee), 1–9. Woodhead Publishing.

21 Brockgreitens, J. and Abbas, A. (2016). Responsive food packaging: recent progress and technological prospects. *Compr. Rev. Food Sci. Food Saf.* 15: 3–15.

22 Fuciños, C., Míguez, M., Cerqueira, M.A. et al. (2015). Functional characterization and antimicrobial efficiency assessment of smart nanohydrogels containing natamycin incorporated into polysaccharide-based films. *Food Bioprocess Technol.* 8: 1430–1441.

23 Cerqueira, M.A., Costa, M.J., Fuciños, C. et al. (2014). Development of active and nanotechnology-based smart edible packaging systems: physical-chemical characterization. *Food Bioprocess Technol.* 7: 1472–1482.

24 Yadav, S.K., Jung, Y.C., Kim, J.H. et al. (2013). Mechanically robust, electrically conductive biocomposite films using antimicrobial chitosan-functionalized graphenes. *Part. Part. Syst. Char.* 30: 721–727.

25 Sundramoorthy, A.K. and Gunasekaran, S. (2014). Applications of graphene in quality assurance and safety of food. *TrAC, Trends Anal. Chem.* 60: 36–53.

26 Kim, H. and Macosko, C.W. (2009). Processing-property relationships of polycarbonate/graphene composites. *Polymer (Guildf.)* 50: 3797–3809.

27 Basu, P.K., Indukuri, D., Keshavan, S. et al. (2014). Graphene-based *E. coli* sensor on flexible acetate sheet. *Sens. Actuators, B* 190: 342–347.

28 Ghaani, M., Cozzolino, C.A., Castelli, G., and Farris, S. (2016). An overview of the intelligent packaging technologies in the food sector. *Trends Food Sci. Technol.* 51: 1–11.

29 Realini, C.E. and Marcos, B. (2014). Active and intelligent packaging systems for a modern society. *Meat Sci.* 98: 404–419. https://doi.org/10.1016/j.meatsci.2014.06.031.

30 Noguera, S.J.C. and Noguera, J.S.C. (2015). Nanotechnology smart sticker for use with perishable foods. US 2015/0056333 A1. https://www.google.com/patents/US20150056333 (accessed 01 February 2021).

31 Suntres, Z.E. and Shek, P.N. (1996). Alleviation of paraquat-induced lung injury by pretreatment with bifunctional liposomes containing alpha-tocopherol and glutathione. *Biochem. Pharmacol.* 52 (10): 1515–1520.

32 (2014). Preparation method of nano TiO_2 powder and method for preparing oxygen gas indicator from nano TiO_2 powder. CN 103641163 A. https://www.google.com/patents/CN103641163A?cl¼en (accessed 01 February 2021).

33 Marsanasco, M., Márquez, A., Wagner, J. et al. (2011). Liposomes as vehicles for vitamins E and C: an alternative to fortify orange juice and offer vitamin C protection after heat treatment. *Food Res. Int.* 44 (9): 3039–3046.

34 Bright, M.A. (2016). NFC/RF mechanism with multiple valid states for detecting an open container, and methods of making and using the same. US 10,579,919 B2. https://www.google.pt/patents/US9519904 (accessed 01 February 2021).

35 Calani, L., Brighenti, F., Bruni, R., and Del Rio, D. (2012). Absorption and metabolism of milk thistle flavanolignans in humans. *Phytomedicine* 20: 40–46.

36 Javed, S., Kohli, K., and Ali, M. (2011). Reassessing bioavailability of silymarin. *Altern. Med. Rev.* 16 (3): 239–249.

37 Mozafari, M.R. and Mortazavi, S.M. (2005). *Nanoliposomes: From Fundamentals to Recent Developments*. Oxford: Trafford Publishing Ltd. ISBN: 1-4120-5545-8.

38 Gregoriadis, G. (1984). Liposome technology. In: *Preparation of Liposomes*, vol. 1, 268. Boca Raton, FL: CRC Press.

39 Gonnet, M., Lethuaut, L., and Boury, F. (2010). New trends in encapsulation of liposoluble vitamins. *J. Controlled Release* 146: 276–290.

40 Kirilenko, V.N. and Gregoriadis, G. (1993). Fat-soluble vitamins in liposomes: studies on incorporation efficiency and bile salt-induced vesicle disintegration. *J. Drug Targeting* 1 (4): 361–368.

41 Guidance for Industry (2012). *Assessing the Effects of Significant Manufacturing Process Changes, Including Emerging Technologies, on the Safety and Regulatory Status of Food Ingredients and Food Contact Substances, Including Food Ingredients that are Color Additives*. US FDA https://www.fda.gov/regulatory-information/search-fda-guidance-documents/guidance-industry-assessing-effects-significant-manufacturing-process-changes-including-emerging.

42 Storma, G., Belliota, S.O., Daemen, T., and Lasic, D.D. (1995). Surface modification of nanoparticles to oppose uptake by the mononuclear phagocyte system. *Adv. Drug Delivery Rev.* 17: 31–48. https://doi.org/10.1016/0169-409X[95]00039-A.

43 Lamprecht, A., Ubrich, N., Yamamoto, H. et al. (2001). Biodegradable nanoparticles for targeted drug delivery in the treatment of inflammatory bowel disease. *J. Pharmacol. Exp. Ther.* 299: 775–781.

44 Solaro, R., Chiellini, F., and Battisti, A. (2010). Targeted delivery of protein drugs by nanocarriers. *Materials* 3: 1928–1980. https://doi.org/10.3390/ma3031928.

45 Hunter, A.C., Elsom, J., Wibroe, P.P., and Moghimi, S.M. (2012). Polymeric particulate technologies for oral drug delivery and targeting: a pathophysiological perspective. *Maturitas* 73: 5–18. https://doi.org/10.1016/j.maturitas.2012.05.014.

46 Reis, P.C., Neufeld, R.J., Ribeiro, A.J., and Veiga, F. (2006). Nanoencapsulation II. Biomedical applications and current status of peptide and protein nanoparticulate delivery systems. *Nanomedicine* 2: 53–65. https://doi.org/10.1016/j.nano.2006.04.0091.

47 Kumari, A., Yadav, S.K., and Yadav, S.C. (2010). Biodegradable polymeric nanoparticles based drug delivery systems. *Colloids Surf., B* 75: 1–18. https://doi.org/10.1016/j.colsurfb.2009.09.001.

48 De Mattos, A.M., Olyaei, A.J., and Bennett, W.M. (2000). Nephrotoxicity of immunosuppressive drugs: long-term consequences and challenges for the future. *Am. J. Kidney Dis.* 35: 333–346. https://doi.org/10.1016/S0272-6386[00]70348-9.

49 Shi, W., Chen, M., Xie, L. et al. (2012). A novel cyclosporine a drug-delivery system for prevention of human corneal rejection after high-risk keratoplasty: a clinical study. *Ophthalmology* 120: 695–702. https://doi.org/10.1016/j.ophtha.2012.09.035.

50 Romero, G.B., Arntjen, A., Keck, C.M., and Müller, R.H. (2016). Amorphous cyclosporin A nanoparticles for enhanced dermal bioavailability. *Int. J. Pharm.* 498: 217–224. https://doi.org/10.1016/j.ijpharm.2015.12.019.

51 El-Shabouri, M.H. (2002). Positively charged nanoparticles for improving the oral bioavailability of cyclosporine-A. *Int. J. Pharm.* 249: 101–108. https://doi.org/10.1016/S0378-5173[02]00461-1.

52 Cotter, P.D., Hill, C., and Ross, R.P. (2005). Bacteriocins: developing innate immunity for food. *Nat. Rev. Microbiol.* 3: 777–788. https://doi.org/10.1038/nrmicro1273.

53 Schnell, N., Entian, K.-D., Schneider, U. et al. (1988). Propeptide sequence of epidermin, a ribosomally synthesized antibiotic with four sulphide-rings. *Nature* 333: 276–278. https://doi.org/10.1038/333276a0.

54 Bierbaum, G. and Sahl, H.G. (2009). Lantibiotics: mode of action, biosynthesis and bioengineering. *Curr. Pharm. Biotechnol.* 10: 2–18. https://doi.org/10.2174/138920109787048616.

55 Mulders, J.W., Boerrigter, I.J., Rollema, H.S. et al. (1991). Identification and characterization of the lantibiotic nisin Z, a natural nisin variant. *Eur. J. Biochem.* 201: 581–584. https://doi.org/10.1111/j.1432-1033.1991.tb16317.x.

56 Laridi, R., Kheadr, E.E., Benech, R.O. et al. (2003). Liposome encapsulated nisin Z: optimization, stability and release during milk fermentation. *Int. Dairy J.* 13: 325–336. https://doi.org/10.1016/S0958-6946[02]00194-2.

57 Abreu, L., Todaro, V., Sattler, P. et al. (2016). Development and characterization of nisin nanoparticles as potential alternative for the recurrent vaginal candidiasis treatment. *AAPS PharmSciTech* 17: 1421–1427. https://doi.org/10.1208/s12249-016-0477-3.

58 Arnold, T.M., Forrest, G.N., and Messmer, K.J. (2007). Polymyxin antibiotics for gram-negative infections. *Am. J. Health Syst. Pharm.* 64: 819–826. https://doi.org/10.2146/ajhp060473.

59 Myint, T., Evans, M.E., Burgess, D.R., and Greenberg, R.N. (2016). Respiratory muscle paralysis associated with colistin, polymyxin B, and muscle relaxants drugs: a case report. *J. Invest. Med. High Impact Case Rep.* 4: 1–4. https://doi.org/10.1177/2324709616638362.

60 Gallardo-Godoy, A., Muldoon, C., Becker, B. et al. (2016). Activity and predicted nephrotoxicity of synthetic antibiotics based on polymyxin B. *J. Med. Chem.* 59: 1068–1077. https://doi.org/10.1021/acs.jmedchem.5b01593.

61 Severino, P., Silveira, E., Loureiro, K. et al. (2017). Antimicrobial activity of polymyxin-loaded solid lipid nanoparticles (PLX-SLN): characterization of physicochemical properties and in vitro efficacy. *Eur. J. Pharm. Sci.* 106: 177–184. https://doi.org/10.1016/j.ejps.2017.05.063.

62 Alipour, M., Halwani, M., Omri, A., and Suntres, Z.E. (2008). Antimicrobial effectiveness of liposomal polymyxin B against resistant gram-negative bacterial strains. *Int. J. Pharm.* 355: 293–298. https://doi.org/10.1016/j.ijpharm.2007.11.035.

63 Liu, C., Kou, Y., Zhang, X. et al. (2018). Strategies and industrial perspectives to improve oral absorption of biological macromolecules. *Expert Opin. Drug Delivery* 15: 223–233. https://doi.org/10.1080/17425247.2017.1395853.

64 Harloff-Helleberg, S., Nielsen, L.H., and Nielsen, H.M. (2017). Animal models for evaluation of oral delivery of biopharmaceuticals. *J. Controlled Release* 268: 57–71. https://doi.org/10.1016/j.jconrel.2017.09.025.

65 Sneh-Edri, H., Likhtenshtein, D., and Stepensky, D. (2011). Intracellular targeting of PLGA nanoparticles encapsulating antigenic peptide to the endoplasmic reticulum of dendritic cells and its effect on antigen cross-presentation in vitro. *Mol. Pharmaceutics* 8: 1266–1275. https://doi.org/10.1021/mp200198c.

66 Nicolosi, D., Scalia, M., Nicolosi, V.M., and Pignatello, R. (2010). Encapsulation in fusogenic liposomes broadens the spectrum of action of vancomycin against gram-negative bacteria. *Int. J. Antimicrob. Agents* 35: 553–558. https://doi.org/10.1016/j. ijantimicag.2010.01.015.

67 Iwasaki, T., Ishibashi, J., Tanaka, H. et al. (2009). Selective cancer cell cytotoxicity of enantiomeric 9-mer peptides derived from beetle defensins depends on negatively charged phosphatidylserine on the cell surface. *Peptides* 30: 660–668. https://doi.org/10.1016/j.peptides.2008.12.019.

68 Teixeira, V., Feio, M.J., and Bastos, M. (2012). Role of lipids in the interaction of antimicrobial peptides with membranes. *Prog. Lipid Res.* 51: 149–177. https://doi.org/10.1016/j.plipres.2011.12.005.

69 Kalpana, S.R., Rashmi, H.B., Rao, N.H., and Ilyas, S.M. (2010). Integrating nanotechnology into agri-food systems research in India: a conceptual framework. *Technol. Forecast. Social Change* 77: 639–648.

70 Kalpana, S.R., Rashmi, H.B., and Jyothi, B. (2011). Research and development perspectives of transgenic cotton: evidence from patent landscape studies. *J. Intellect. Prop. Rights* 16: 139–153.

71 Li, J.J., Muralikrishnan, S., Ng, C.-T. et al. (2010). Nanoparticle-induced pulmonary toxicity. *Exp. Biol. Med.* 235: 1025–1033.

5

Nanotechnology: Improvement in Agricultural Productivity

CHAPTER MENU

5.1 Nanoparticle Biosynthesis and Use in Agriculture

For the synthesis of nanoparticles using toxic chemicals, several chemical methods are available, so the need for time is to use environmentally benign, greener, and eco-friendly routes. Researchers are looking forward to synthesizing nanoparticles from different biological entities such as bacteria, fungi, higher plants, actinomycetes, and viruses. They can also reduce the salts to the corresponding nanoparticles. For the synthesis of nanoparticles, numerous biological sources have been used for precision farming in agriculture [1]. Some of them are as follows: silver nanoparticles, zinc oxide (ZnO) nanoparticles, and titanium dioxide nanoparticles.

5.1.1 Silver Nanoparticles

In silver nanoparticles, the high surface area and the fraction of surface atoms are present and have a significant antimicrobial impact compared to traditional silver [2]. The antimicrobial properties of silver nanoparticles [3–6] are used against many human

Nano-Technological Intervention in Agricultural Productivity, First Edition.
Javid A. Parray, Mohammad Yaseen Mir and Nowsheen Shameem.

pathogens. There is a need to examine the full potential for crop defence. Consequently, the antimicrobial properties of silver nanoparticles are increasingly being used to combat and control plant diseases [7].

Silver nanoparticles have been used as pesticides to reduce the danger of pests coming from plants. Silver nanoparticles can be synthesized using mechanical, chemical, and biological processes. Bio-based approaches are, however, commonly used to escape extreme environments and hazardous chemical requirements. Numerous experiments have effectively synthesized silver nanoparticles as a single phase from different sources (plants, bacteria, fungi, etc.) and are eco-friendly. These silver nanoparticles have been used to destroy harmful microorganisms in plants.

Biological synthesis of silver nanoparticles ranging from 6 to 38 nm from *Raphanus sativus* var. aegyptiacus has been documented. In a laboratory experiment, the snails and soil matrix's exposure to silver nanoparticles decreased the behaviour and viability of the terrestrial snails (20% of the silver nanoparticle-treated snails died) and the frequency of the fungal population in the surrounding soil [8]. Spherical nanoparticles of silver form in a size range of 10–20 nm with *Serratia* sp. culture BHU-S4 and its successful implementation have been experimented with in the treatment of spot blotch disease by *Bipolaris sorokiniana* in wheat have been studied.

Silver nanoparticles demonstrated good antifungal activity against the wheat spot blotch pathogen *Bipolaris sorokiniana* [7]. The impact of silver nanoparticles on fenugreek seeds (*Trigonella foenum-graecum*) was also achieved [9]. The results show good germination of the seeds (76.11%), germination speed (4.102), root length (76.94 mm), fresh root weight (2.783), and root dry weight (1.204) at a concentration of 10 μg/ml. These findings revealed the potential for significantly improving the germination potential, mean germination time, seed germination index, seed vigour index, seed fresh weight, and dry weight of silver nanoparticles.

5.1.2 Zinc Oxide Nanoparticles

Zinc deficiency is the most common micronutrient problem that adversely affects the agricultural production of alkaline soils with calcium carbonate ($CaCo_3$) [10]. Calcium carbonate soils are a significant source of cultivation in the world's arid or Mediterranean environments [11]. Alkaline pH, which decreases zinc solubility, and high calcium carbonate ($CaCO_3$) content, which can absorb and precipitate zinc, are the parameters that restrict the availability of zinc to plants in calcium carbonate soils [12, 13]. ZnO and zinc sulfates ($ZnSO_4{\cdot}H_2O$) or ($ZnSO_4{\cdot}7H_2O$) are commonly used as zinc fertilizers to correct the deficiency of zinc in soils [14]. However, because of the unavailability of zinc for plants, their applications as fertilizers are restricted. Because of their higher reactivity than micron- or millimetre-sized zinc particles present in bulk, ZnO nanoparticles can solve this problem by providing plants with a more soluble and usable form of zinc. The use of ZnO nanoparticles as zinc fertilizers can increase zinc dissolution and its bioavailability in calcium carbonate soils.

The key mechanism for transporting zinc from fertilizer to plant roots following the dissolution process is the diffusion of dissolved zinc [15]. Nanoparticles of ZnO have shown far better antimicrobial activity than large particles of zinc. The small size of nanoparticles below 100 nm and the high surface-to-volume ratio of nanoparticles allow better bacterial interaction [16]. ZnO nanoparticles can induce reactive oxygen species (ROS) production, which can lead to cell death when the cell's antioxidant ability is exceeded [17–21]. On exposure to silver and ZnO engineered nanoparticles on duckweed (*Spirodela punctata*), the production of reactive nitrogen species (RNS) and hydrogen peroxide (H_2O_2) indicated that the toxicity of silver and ZnO nanoparticles was primarily caused by both the particulates and ionic forms [22]. Zinc nanoparticles have been shown to induce free radical formation in wheat, leading to increased malondialdehyde levels and decreased glutathione levels [23] and reduced levels of chlorophyll [24]. By chemical and biological methods, ZnO nanoparticles can be synthesized. As chemical processes need toxic chemicals, natural methods are becoming common. It is cost-effective and eco-friendly to synthesize ZnO nanoparticles from plants. Plant leaf extract has been widely used for the synthesis of nanoparticles with ZnO. Suitable concentrations of either zinc sulfate heptahydrate ($ZnSO_4{\cdot}7H_2O$) or zinc acetate dehydrate (Zn $(CH_3COO)_2{\cdot}2H_2O$) are dissolved in water to synthesize ZnO nanoparticles. Using solvents such as water, ethanol, or methanol, plant leaf extract may be prepared. ZnO nanoparticles are synthesized by combining plant extract and zinc sulfate heptahydrate or zinc acetate dehydrate solution at the desired pH.

ZnO nanoparticles have been tested as bactericidal and fungicidal in the laboratory. ZnO nanoparticles have been synthesized using *Moringa oleifera* leaf extract in the 16–20 nm range, and antimicrobial activity has been evaluated against bacterial strains such as *Staphylococcus aureus*, *Bacillus subtilis*, *Pseudomonas aeruginosa*, *Proteus mirabilis*, *Escherichia coli*, and fungal strains such as *Candida albicans* and *Candida tropicalis*. In *S. aureus*, the highest inhibition zone was observed (23.8 ± 0.76) relative to the others [25]. Spherical and hexagonal ZnO nanoparticles from *Parthenium hysterophorus* L. have been synthesized by an inexpensive, eco-friendly, and simple method using different concentrations of 50% and 25% of Parthenium leaf extracts with size 27 ± 5 and 84 ± 2 nm, respectively. These ZnO nanoparticles were explored for the size-dependent antifungal activity against plant fungal pathogens, i.e. *Aspergillus flavus* and *Aspergillus niger*. A maximum inhibition zone was observed for 27 ± 5 nm size ZnO nanoparticles against *A. flavus* and *A. niger*. Parthenium-mediated ZnO nanoparticles proved to be good antifungal agents and environment friendly [26]. The ZnO nanoparticles having a spherical shape with an average size of 23–57 nm were prepared from zinc acetate and sodium hydroxide obtained from leaves of *Catharanthus roseus*. The synthesized ZnO nanoparticles were evaluated for antibacterial activity against Gram-negative bacteria *E. coli* (ATCC 25922), *P. aeruginosa* (ATCC 15442), Gram-positive *S. aureus* (ATCC 6538), and *Bacillus thuringiensis* (ATCC 10792). Resistance to ZnO nanoparticles was shown by *B. thuringiensis*, followed by *E. coli*, while *P. aeruginosa* was more susceptible. This research concluded that ZnO nanoparticles could beused against *P. aeruginosa* as antibacterial formulations [27].

5.1.3 Titanium Dioxide (TiO_2) Nanoparticles

Titanium is a tough, glorious, and corrosion-resistant metal. A common photocatalyst used in pigment making is the titanium dioxide compound [28]. Titanium stimulates further production of carbohydrates, encouraging the growth and photosynthesis rate in plants [29–31]. Catalytic photographic activity for pesticide degradation has been demonstrated by titanium dioxide [32]. The photocatalyst property of titanium dioxide has plant safety applications because it does not form toxic and hazardous compounds and has excellent potential for pathogen disinfection. By dye doping and other acceptable methods, scientists are trying to increase the phytopathogenic disinfection performance of titanium dioxide thin films [33]. For the synthesis of titanium dioxide nanoparticles, plants are also the primary choice of researchers. Spherical, clustered nanoparticles of titanium dioxide with a mean size of 32.58 nm have been synthesized from the aqueous leaf extract of *Psidium guajava* [34]. These nanoparticles were tested against *Aeromonas hydrophila* (MTCC-1739), *P. mirabilis* (MTCC-442), *E. coli* (MTCC-1677), *S. aureus* (MTCC-3160), and *P. aeruginosa* (MTCC-4030). The maximum inhibition zone was observed against *S. aureus* (25 mm) and *E. coli* (23 mm) when titanium dioxide nanoparticles were used at a 20 μg/ml concentration. Synthesized TiO_2 nanoparticles demonstrated enhanced antibacterial activity compared to tetracycline, the typical antibiotic disk, which decreased the chances of bacterial species developing antibiotic resistance. In contrast with ascorbic acid, the aqueous extract of plants and titanium dioxide nanoparticles synthesized from them had the best antioxidant activity. The synthesis of spherical clusters of very polydispersed titanium dioxide nanoparticles with a size range from 36 to 68 nm by *Eclipta prostrata* leaf extract was successfully carried out [35].

5.2 Nanorobots

This section introduces a groundbreaking approach to hydraulic conductivity evaluation, taking nanorobots into account as a framework capable of opening up new perspectives in the field of hydrological monitoring. The application of nanorobots for agricultural purposes and monitoring water and soil qualities may result in impressive impact towards environmental control and decreasing the damages caused by pollution to many different natural species. It is predicted that applications of nanorobots have remarkable possibilities. Recent developments in the field of biomolecular computing [36, 37] have demonstrated the feasibility of processing logic tasks by bio-computers [38], which is a promising first step to enable future nanoprocessors with increased complexity. Studies targeted at building biosensors [39, 40] and nanokinetic devices [41], required to allow for nanorobotic operation and locomotion, have been advancing recently as well.

We have gained insights into the hydraulic conductivity of broken and karstic rocks over the past 15 years by incorporating particles of varying size, charge, and chemical composition into a flow field and tracking the space and time breakthrough particles. The hydraulic aperture of the smallest throats in a flow path can be inferred from this data. Therefore, we can use nanorobots to expand this definition to porous media [42]. A computational approach for investigating nanorobots' manufacturing design [43] is defined to allow a better understanding of hydraulic conductivity tools. Nanorobots use chemical gradients and electromagnetic sensing to solve integrated hydraulic conductivity calculations over short distances along complex flow paths. Such a method of acquiring knowledge is beneficial in identifying geological characteristics, which are of utmost importance when agricultural management or environmental disasters occur, requiring effective decision-making in a short period or improving productivity in certain industrial activities. In 2006, nanotechnology-based environmental applications' overall demand was estimated at US$374.9 million, reaching more than $6.1 billion by 2010 [44]. Geophysicists have held out hope of ways to describe hydraulic conductivity distribution with new analytical and detection methods. Yet, we are a little closer to the illumination of this Holy Grail than we were 40 years ago. The limitation of geophysical methods falls into two categories. First, there is a limitation of direct measurement on the size of the features we are looking for (pore throats). These small features require a short wavelength and thus high energy for resolution. Unfortunately, life is rapidly dissipated in travel through earth material, generally resulting in degraded resolution at the desired scales. This wavelength/energy constraint is fundamental and cannot be overcome. Second, because geophysicists cannot directly measure the feature of interest, they measure a surrogate of pore-throat sizes. Typical geophysical methods measure the properties of waves, density, or electrical conductivity used to generate lithologic information.

The ways to achieve electromagnetic sensing devices embedded into the nanorobot are described. When combined with a nano-positioning system for a four-dimensional reference location, it could also store data from these sensors in memory that could be downloaded after capturing the nanobots at some distance along the flow field accessed by a signal from the surface, defining this way preferential flow paths. Such an approach could be used even for tracking distinct materials – oil, uranium, or gold, to quote a few. The nanorobots may observe hardware feasibility because of the most recent advances in a broad range of manufacturing techniques. Within the miniaturization trends, we can also cite some examples such as very large-scale integration (VLSI) chips, including the latest technology-based complementary metal oxide semiconductor (CMOS) [36], which could be seen as one possible way to generate embedded control computation on molecular machines shortly [45]. In the meantime, while these manufacturing methodologies can advance gradually, the use of computer techniques and virtual reality could contribute to the transducer investigation process [46]. Therefore, this work aims to outline ways to produce nanorobotic on-chip systems to prepare their use for potential applications that will concern agricultural, industrial, and environmental issues.

5.2.1 Environment Monitoring

Applications in remediation, protection, maintenance, and enhancement [44] may be listed as critical areas for future environmental management. SoC (System-on-Chip) innovations could allow hardware manufacturing using CMOS and other related technologies in the coming few years. It could successfully apply the chemical, electromagnetic, and thermal parameters to achieve nanorobots' reasonable control to trigger its actuation upon predefined parameters in open workspaces. Such an approach ensures feasible manufacturing perspectives and enables accurate control activation. As part of our ongoing development and control system design validation projects, we use computer-aided design software, including the critical parameters needed for the investigation at micro-nanoscopic levels. The simulation included a three-dimensional (3D) real-time analysis visualization using the Nanorobot Control Design (NCD) programme to perform a predefined series of tasks based on environment sensing and trigger behaviours (Figure 5.1).

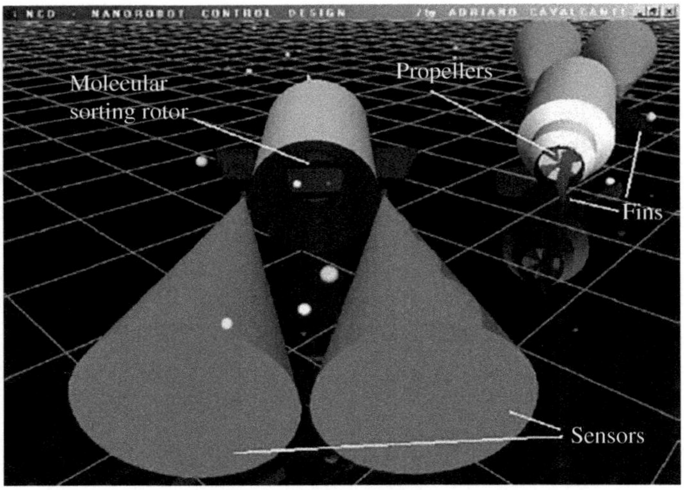

Figure 5.1 Nanorobot 3D design. Source: Boehm [44].

Most of the planet Earth's surfaces are filled by water. The hydrological cycle is of particular significance and direct effect with substantial implications on all humanity and the world's life systems. Only the ocean accounts for 70% of the world's surface area. Excellent or poor hydrological resource management can now have drastic and catastrophic effects on agricultural production over the next few years. Wireless sensors with dispersed robotic agents in selected areas can be one of the most efficient ways to track large areas of natural resources, especially human-driven boats monitoring water resources 24 hours under poor or bad conditions [47]. The benefits of using nanorobots for environmental tasks are apparent: more control over the measurement of microorganisms, better detection of chemical contaminants, and enhanced water temperature control are some of the few positive aspects. Because of more effective monitoring of environmental changes, technology may save many lives from devastating storms or natural disasters. It is possible to expect enormous benefits from nanotechnology and the miniaturization of electronics. The use of nanotubes and nanowires has shown a wide variety of possibilities for use in the manufacture of better nanoscale sensors and actuators. The manufacturing design of nanorobots with integrated embedded nanoelectronics and circuits [46] may be facilitated by these innovations connected with 3D computational simulation. The use of nanorobots for environmental monitoring may be easier to assemble than biomedical nanorobots by introducing the first set of nanomachines consisting of harmonically integrated nanoelectromechanical system (NEMS) and microelectromechanical system (MEMS) devices. It could be possible because nanorobots' sizes are not so stringent for hydrological monitoring as in biomedical problems, where the dimensions of nanorobots in most cases could not reach 3 μm in diameter. Such innovations of nanorobots in environmental matters can also help medicine, mainly because research into the manufacture, control, and operation of nanorobots in natural environments could facilitate the first nanorobots tailored for biomedical applications at the same time.

5.2.2 Nanorobot Sensors

The nanorobot's proposed prototype design must fit with the necessary devices to track its functional workspace's most critical aspects. Depending on the situation, when monitoring hydrological tools, temperature, chemical concentration in the water, and electrical conductivity are some of the essential parameters. It is expected that geographically dispersed nanorobotic teams will open up new possibilities for agricultural and environmental applications with a broader range of information not seen at all. Computer processing, energy supply, and data transfer capabilities can be addressed for these purposes by embedded integrated circuits, using developments in technology derived from VLSI design [48]. CMOS VLSI production using deep ultraviolet lithography offers high accuracy and a massively commercial way of producing nanoelectronics. The CMOS industry will successfully build the route to allow nanorobots' assembly. The everyday use of nanophotonic and nanotubes will further accelerate the actual resolution levels from 248 to 157 nm devices [49]. Hardware description language (VHDL) has become the most common technique used in the integrated circuit industry to validate designs and achieve efficient implementation [50].

5.2.3 Pollutant and Chemical Detection

In the past 15 years, the production of silicon- and chemical-based sensor arrays using the two-level device architecture hierarchy has been successfully carried out [51, 52]. The technology ranges from air detection to water element pattern recognition by embedded software programming in the automotive and chemical industries. It can reduce the high energy demand costs for data transmission and circuit operation by about 60% by using nanowire [53]. For applications concerning chemical changes, both in environmental care and in biomedical applications, CMOS sensors using nanowires as a material for circuit assembly may achieve maximum performance. For CMOS functionality, sensors with suspended arrays of nanowires installed into silicon circuits can dramatically reduce self-heating and thermal coupling [54]. Using integrated circuits, nanometre chemical sensors will produce enormous low-cost profits for massively manufacturing commercial devices with a wide range of applications in medical, automotive, and environmental problems and much more. Factors such as low consumption of energy and high sensitivity are among the benefits of nanosensors. The manufacturing array method of nanosensors can use electrofluidic alignment as a multielement to achieve an integrated CMOS circuit assembly [53]. To allow cross-sectional drive transistors for signal processing circuit read-out, passive and buried electrodes should be used. To prevent the loss of processed signals, the passive and buried aligned electrodes must be electrically separated. Control feedback to switch the electronic sensor between active or switched off detection is necessary for monitoring operations. For integrated sensors, the nanowire is intended to achieve breakthroughs in high-speed sensing and control technologies. The range for CMOS operation is $\sim$190 MHz for chemical sensor devices. The most suitable route for saving the nanorobot's energy when the sensor is in the process of data transmission control can use sample techniques for operational conditions with a time interval of 80 ns [53].

Quantum mechanical tunnelling for the operation of thin oxide gates and subthreshold slope [55] are some of the constraints for improving actual CMOS and metal–oxide–semiconductor field-effect transistor (MOSFET) methodologies. The semiconductor branch has moved forward to keep the circuit moving forward, overcoming any expectations. With new materials to meet the increasing demand for highly complex VLSIs, smaller channel lengths and lower voltage circuits for higher performance are being achieved. To beat self-heating and improve performance [56], new materials such as strained channel with a relaxed SiGe layer are cited. Recent developments in 3D circuits and double-gate FinFETs have achieved astonishing results and should be further improved according to the semiconductor roadmap. Silicon-on-insulator (SOI) technology has been used to manufacture high-performance logic sub 90 nm circuits [57, 58] to advance different production techniques. Circuit design approaches have been successfully shown to solve bipolar and hysteretic variations based on SOI structures [56]. Overall, as a real breakthrough in technology devices, it transforms feasible 90 and 65 nm CMOS devices into products that can be strategically used.

5.2.4 Metal Identification

There are substantial application possibilities for sensors for the identification of different metals. Electromagnetic sensors could be used in various cases: massive circuit board production analysis, industrial assessment of the properties of magnetic materials, naval operation, and medical diagnosis or even quality control for certain food industry products [59, 60]. These sensors have also been widely used to classify tiny or large parts of rocks and related materials. Electromagnetic induction sensors are the most appropriate means of detecting metal and low-metal content [61]. It is used to detect metal fragments from different depths and different environmental conditions, with distances varying according to the object's size and characteristics. Today, the number of mines classified worldwide is around 100 million [62]. Active electromagnetic sensors can enable a wider variety of identification patterns compared to passive sensors [63]. It should easily apply to the classification of both ferromagnetic and non-magnetic metallic objects and, in certain circumstances, to non-metallic objects [63]. It works with electromagnetic pulses with a signal proportional to the magnetic resonance change rate, which is affected by the tolerance of the orientation and angular dependence on the incident beam above the target to be identified. It will reflect the distance, shape, electrical conductivity, and magnetic permeability in the signal received back by the sensor and should consider these aspects for spectral recognition [63]. The use of electromagnetic resonance helps short-distance identification when considering the energy concerns of sensors on a nanochip. This can be done successfully by teams of nanorobots with integrated CMOS technology, which could be used in an unprecedented way to coordinate short-range sensing. Multifrequency calibration is required to modulate the electromagnetic sensor operation to suit different types of targets. An essential rule for the production of CMOS sensor devices is also expected to play a crucial role in achieving higher photonic technology performance on electromagnetic sensors [64]. It significantly improves the functionality of the sensor and precision, allowing subcomponents to be better matched. This is done using lithography for chip implementation without costly active methods because of intrinsic photonic properties [65]. Therefore, high-precision nanoelectronics can be manufactured using photonic properties using e-beam lithography with wavelengths down to 300 nm [66, 67].

Nanophotonic SOI UV lithography could be used to fabricate high-performance CMOS sensors, which permits precision on assembly process range 1–10 nm scales [68]. Photonic wires are more appropriate for such process because they are not periodic having lower dispersion [49].

5.2.5 Nanorobot Data Transmission

The use of the sensor to transmit environmental and agricultural resource monitoring data has shown significant benefits. To classify inventory livestock and probably monitor disease spread, radio frequency identification (RFID) sensors have been used [69]. Environmentalists and the fishing industry have used a small RFID integrated circuit to analyse better the salmon population's decline in recent years with its potential relationship to global warming and changes in the environment and an automatic device to capture salmon more efficiently [70]. Automated robotics will bring significant changes in 24-hour monitoring with higher precision and lower costs when dealing with underwater environments. Nanorobots are expected to have a considerable influence on the collection of data in such scenery. The use of nanorobots integrated sensor systems will directly support not only hydrological search and analytical environmental conclusion reports but also the gas and oil industry operating procedures. It may serve as a mobile device for tracking vast areas underwater or in harsh environments.

Thus, recurrence and maintaining operations could become more secure, automatic, and with lower costs through the use of smart sensors to help identify what repair should be urgently necessary on underwater pipelines and pumping types of equipment [71]. Prevision of low-cost polymer electronics for tiny magnetic sensors and transducer chips is expected to achieve low costs as 1 cent for the circuit, making its use even more attractive [72].

Electromagnetic fields combined with passive sonar signature techniques should be the best data acquisition approach for monitoring surrounding underwater environments [63, 73, 74]. The teams of nanorobots should be equipped with CMOS sensors [75]. It could use CMOS with submicron SoC design for low power consumption with nanorobots communicating through sonar sensors [76]. For the nanorobot, active sonar communication frequencies may reach up to 20 μW at 8 Hz as resonance rates and 3 V supply [77]. Mixing static and mobile sensors for data transmission is a useful and widely accepted approach for monitoring environmental and hydrological resources [47]. Thus, strategically positioned static sensors for acquiring wireless data transmission from mobile nanorobots comprise an excellent choice to patrol and monitor predefined patterns from some target area. To accomplish that, acoustic micron sensors may be exchanging communication and strategic data information should be transmitted when some new event was registered from nanorobots as mobile sensor in the specified region.

For communication in liquid workspaces, we can cite acoustic, light, and chemical signals as potential solutions for communication and data transmission depending on the purpose in mind and the range of distances [45, 47]. For some teamwork coordination, chemical communication is constructive for nearby communication between nanorobots [39]. On the other hand, acoustic communication, in contrast to light communication methods, is more suitable for long-distance transmission and detection with low energy consumption [45].

Although optical communication enables faster data transmission rates [47], its energy demand does not make it suitable for nanorobots. The best solution for reading and writing data in underwater workspaces is integrating integrated sensors for data transmission. An electromagnetic reader is used in our design to launch waves and detect the current status of nanorobots in the distribution field. This transponder system transmits magnetic signatures to the nanorobot-embedded passive CMOS sensors that allow data to be sent and received through electromagnetic fields. The data monitoring nanorobots convert the wave propagation produced through a well-defined protocol by the emitting devices. According to the last collection of events reported in pattern arrays, the wave's resonance [72] will mirror information back. Data transmission ranges are feasible for nanorobot passive data transfer of ~4.5 kHz frequency with approximately 22 μs delays. While such a passive mechanism can be used to collect data from nanorobots, sonar communication is to be used for active communication between nanorobots to perform synchronised behaviours due to certain unique group tasks that need to be performed.

5.2.6 Nanorobot System on Nanotechnology Chip

Advances in nanotechnology can allow the development of short-term nanosensors and actuators. Indeed, with the introduction of nanotransistors and integrated circuits, this is happening. Hybrid techniques using nanotubes, photonics, and mesoscopic nanowires as elements for design are potential breakthroughs for it. We do not have such a strict parameter size restriction for hydrology monitoring. Photolithographic patterning of the insulator layer under the MEMS bridge was used as a first attempt towards the potential assembly of nanoscopic components for prototyping a CMOS. Followed by deposition of 1500 Å silicon nitride and acetone lift-off, the g_0 gap dimension is established by a deposit of 1.5 μm sacrificial layer AZ5214E photoresist [78]. The pattering of the windows connected to the columns of the bridge is then finished. To avoid damage to the photoresist caused by the gold bath solution during the structures' electroforming, hard baking is necessary [78]. The electroformation of gold posts of 1.5 μm is included. It included that the silicon nitride layer isolation layer was in the CMOS-built sensor [78].

As a first study case, this manufacturing process CMOS was adopted. Further studies will lead us to downgrade sizes with more precise requirements on the use of nanotubes and advanced photonic materials to achieve integrated circuits of high precision and efficiency. In order to achieve a full nanorobot assembly based on dynamic capability behaviour, as discussed, different sensors should be achieved both in design simulation and manufacturing implementation. VLSI and lithography, together with precision verification methods such as SoC and VHDL, have made impressive breakthroughs in electronic development.

5.3 Natural Nanostructures in Food

In plant and animal tissues, nanostructures are the natural building blocks of a lot of multiscale organizations. This protein, lipid, or polysaccharide nanostructures, alone or mixed, also in combination with inorganic molecules, play a significant role in providing biological materials with specific properties, such as mechanical properties such as spider silk or nacre, energy storage properties such as plant oleosomes, or transport properties such as high-density human lipoprotein. These nanostructures result from the bottom-up assembly of molecules via hydrophobic or electrostatic interactions or covalent links depending on their chemical nature and environmental conditions. There are several distinct naturally occurring particles in food that fall into the category of nanoparticles due to their size in the range of several nanometers, such as casein micelles or fat globules in milk. Owing to these natural structures' large size distribution, only a fraction is actually at a nanometric size (<100 nm). Because of thermomechanical conditions and physicochemical environments, nanoparticles in food come directly from the plant or raw animal products used and may also be produced during processing. Since 10 years, the determination of structural changes occurring during processing and the knowledge of the interaction mechanisms have allowed better understanding of the preservation and alteration of the native nanoparticles and the creation of new ones. Nanotechnology provides new ways of controlling and structuring food with greater functionality and value. Nanotechnology in food science is now becoming an active research field, especially with functional foods.

5.3.1 Protein-Based Nanostructures

Proteins themselves are nanoparticles that can self-assemble into, ordered or amorphous, larger structures. The final architecture's shape and size result from a coexistence of repulsive long-range forces and attractive short-range forces between the protein building blocks by changing the balance between these various forces, a large diversity of structures. Either covalent or noncovalent interactions are generally involved in these assemblies. Depending on the strength and number of interactions per proteins, varying lifetimes may occur. The thermodynamic equilibrium of these assemblies is highly dependent on several factors: composition and concentration of proteins, solvent conditions such as pH, ionic strength, ion existence, and physical conditions such as temperature, local pressure, and so on. A powerful technique to construct complex nanostructures is the bottom-up method of molecular self-assembly that exists in nature. Proteins can spontaneously fold and assemble into structurally well-defined and functional nano-scaled structures [79] throughout a range of intra- (folding) or inter-molecular noncovalent interactions such as hydrophobic effect, hydrogen bonding, electrostatic interaction, and metal binding. To create protein nanostructures with various conformations according to environmental stimuli, nature investigates the complex and adaptive nature of noncovalent, reversible interactions. As such, nanoparticles with a size varying from one to several 100 nm are found in natural foods [80, 81]. Proteins from foods widely consumed as milk and eggs are illustrated in this chapter. Complex biological macromolecules are proteins. Depending on the sequence of amino acids, they form β-sheet or α-helix or non-structured coiled domains; proteins are distinguished from each other by their covalent-linked backbone of amino acids, and these domains interact by noncovalent interactions, leading to tertiary structure (globular shape, fibrous shape, and intrinsically unfolded conformation) and provide their biological proteins. Multiple proteins have a quaternary structure, apart from these three structural stages, resulting from the combination of two or more equivalent or different amino acid chains (protein subunits). A large number of conformations fluctuate between the overall protein conformation in solutions. As the lowest energy structure or highest probability structure, the native protein structure is defined. It is known that proteins are zwitterions, and they contain both positive and negative payments in a specific proportion. Therefore, proteins are weak polyelectrolytes, whose charging sign and density show a firm reliance on physicochemical conditions. Proteins are stable in solution away from their isoelectric point (pI), where the net charge is zero because the protein molecules carry charges of the same sign and repel each other. In general, protein sizes, particularly globular ones, are in the order of a few nanometres, which is an example of naturally occurring nanoparticles [80].

In comparison to intrinsically unfolded proteins, because of the residual charged patches on the protein surface that counterbalance attractive short-range interactions, numerous globular proteins are also stably close to their pI at low or medium ionic strength. In some conditions (elevated temperatures, presence of denaturants, etc.), the native protein unfolds (loss of its native conformation) or denatures. Each protein (except intrinsically unfolded proteins) has a denaturation temperature dependent on the solvent conditions (pH, ionic strength, and dielectric constant). Upon denaturation, the hydrodynamic size, flexibility, and reactivity of the proteins increase because of the exposure of reactive groups to the proteins' surface. This denaturation may lead to protein aggregation and to some extent loss of solubility [79]. As they are naturally present in food, this chapter focuses on explaining many food proteins' structure.

In some cases, we may also discuss the spontaneously formed new nanostructures that can be created following individual proteins' interaction with naturally coexisting bioactive molecules under native conditions. Most of the food proteins obtained from animal or plant origin are globular structures. Some exceptions, such as milk-based caseins, exhibit complex colloidal systems called casein micelles and myofibrillary proteins' fibrous structure with a meat tissue structuring role. Proteins listed below are primarily milk (β-lactoglobulin [β-Lg], α-lactalbumin [α-La], bovine serum albumin [BSA], lactoferrin, and caseins) or white eggs (ovalbumin, lysozyme, and ovotransferrin) because of their high incidence in food [82, 83]. Food proteins exist primarily in the monomeric form, but they can self-assemble into oligomers or aggregates under some unique conditions. The examples of protein nanostructures present in foods are discussed in the following sections.

5.3.1.1 β-Lactoglobulin

Bovine β-lactoglobulin (β-Lg) is a 162 amino acid globular protein with a molar mass of 18.3 kDa and an isoelectric point of approximately 5.2. The native β-Lg monomer's length is around 3.6 nm, and its molecular structure is well developed. β-Lg has 10–15% α-helix, 43% β-sheet, and 47% unordered structures, including β-turn. Its network comprises nine β-sheets (in yellow) that, as determined by X-ray crystallography, are arranged to form a calyx and a C-terminal α-helix [83]. In the sensitivity of β-Lg towards denaturation, two disulfide bonds play an essential role. There is also one free sulfhydryl group in bovine β-Lg, which is buried within the protein structure and is thus non-reactive under physiological conditions but plays an essential role in stabilizing the tertiary structure of the protein. Native β-Lg exists as a stable noncovalent dimer in the pH range of 5.5–7.5 at room temperature. Still, its oligomerization depends on the medium conditions that lead to fractal aggregates that rely on the fractality's ionic strength. The screening of protein charges by the addition of NaCl contributes to an improvement in aggregate size and protein density. For example, under other pH values, below pH 5.5, β-Lg self-assembles into higher ordered oligomers (tetramers) or dissociate into monomers. It can shape Fibrillar structures at pH 2. The precise biological role of β-Lg is not understood so far. However, it belongs to the lipocalin superfamily, sharing the typical β-barrel calyx's structural characteristic as an ideal binding site for hydrophobic ligands. Lipocalins are typically transporters of small molecules that are hydrophobic. Consequently, a possible biological feature is the role of β-Lg in the defence and transport of small bioactive ligands [83].

5.3.1.2 Serum Albumin

Serum albumin is a protein that is well developed, abundant, and multifunctional [84]. BSA is a helical protein with a 582 amino acid polypeptide chain, a molar mass of 66.4 kD, and a pI of 4.9. The 3D BSA monomeric nanoparticle structure is an ellipsoid with a radius of approximately 3.5 nm. BSA comprises 1 group of sulfhydryl and 17 bonds of disulfide close to each other along the polypeptide chain, strongly stabilizing the protein structure. For this function, BSA is commonly used for calibration purposes as the default protein. BSA is also widely used as a model to research hydrophobic ligands' interactions since on its surface, and it exhibits many binding sites [84]. Its property and ability to bind fatty acids and lipids have been correlated with biological activity. BSA is the most abundant protein in blood plasma to date and can carry seven fatty acid molecules, each protein molecule. In vitro studies have indicated that the protein plays a role in mediating the oxidation of lipids.

5.3.1.3 α-Lactalbumin and Lysozyme (Lys)

Hen egg white lysozyme (Lys) and milk **α**-lactalbumin (**α**-La) exhibit exceptional near homology in sequence and tertiary structures but with many biochemical, biophysical, and biological differences [85]. Of the 123 amino acid residues of **α**-La, 54 are identical to the corresponding residues in lysozyme, and a further 23 are structurally similar. **α**-La consists of 123 amino acids and has a 14.2 kDa molecular weight and a 2.01 nm radius. It is an acidic protein between 4.2 and 4.5 with an isoelectric point. **α**-La consists of 26% α-helix, 14% β-sheet, and 60% of the unordered structure. **α**-La contains eight residues of cysteine, all of which are involved in disulfide bonds that stabilize the protein's tertiary structure. Besides, **α**-La precisely binds one high-affinity calcium ion per molecule (K_a = 108 M^{-1}) into a powerfully negatively charged pocket containing four aspartate residues. The decrease in pH contributes to developing the extremely heat-sensitive, calcium-free apo form of **α**-La. The temperature of the denaturation of apo α-La is about 30 °C, while the native holo **α**-La is stable up to around 60 °C. Calcium pocket of α-La can also bind other divalents such as Zn^{2+}, Mg^{2+}, and Mn^{2+}. As one of the two components of lactose synthase, which catalyses the final step in lactose biosynthesis in the lactating mammary gland, **α**-La performs an essential role mammary secretory cells. Other studies have shown that certain conformational types of **α**-La can induce tumour cell apoptosis, implying that this protein has many essential biological functions [85].

Egg white lysozyme (a C-type family without a particular metal-binding site) is a globular protein composed of 129 amino acids (14.3 kDa) with a radius of approximately 2.1 nm. Unlike al-La, Lys is rich in lysine and arginine residues, resulting in an unmistakable essential character (pI 10.7). Like α-La, Lys is a protein consisting of two domains (domains α and β) linked together by a long helix–loop–helix (residues 87–114). The secondary structure of the protein is formed by 39% α-helix gathered mainly in the α-domain and 11% β-sheet involved in a three-strand antiparallel β-sheet that constitutes, with some helices, the β-domain. Lys has eight cysteine residues, all involved in intramolecular disulfide bonds, making the protein compact and stable [85]. It is noticed that under specific conditions of protein concentration, temperature, and ionic strength, Lys forms short nanoclusters of 10–20 nm. These nanoclusters result from short-range attractions, leading to surface energy reduction upon cluster formation and long-range repulsions that increase the clusters' Coulomb energy and limit their growth.

5.3.1.4 Ovalbumin and Avidin

The main protein in egg white is ovalbumin (Ova) and is one of the first proteins isolated in pure form [86]. Ovalbumin is a 385 amino acid phosphoglycoprotein and has a molar mass of approximately 45 kDa and a radius of approximately 3.05 nm. Its amino acid sequence contains about 50% hydrophobic residues and about 33% charged, mainly acidic, residues, giving the protein a pI of 4.5. Six cysteine residues form the series, two of which are involved in a disulfide bond, and an acetylated glycine is the N-terminal residue of the protein. Ovalbumin has a single site for glycosylation and two areas for phosphorylation. As determined by X-ray crystallography, ovalbumin contains 32% β-sheets and 30% β-helix. Some evidence indicates that, depending on protein concentration and pH, ovalbumin molecules self-assemble. At a concentration lower than about 0.1%, ovalbumin is monomeric and forms oligomers (dimer, trimer, and tetramer) at higher protein concentrations; when pH decreases from a neutral value to the protein pI, this association behaviour is preferred. While ovalbumin belongs to the serpin superfamily, it is not well described in its exact biology. Ovalbumin has been proposed to play a role in the development of chick embryos.

Avidin is present in the white eggs of reptiles, insects, and amphibians. It is understood that chicken avidin is an essential protein for widespread biotechnological application forms [87]. The avidin molecule is a noncovalent homotetrameric glycoprotein with a radius of approximately 3.4 nm and a pI of 9.5. There are 128 amino acid residues in each monomer, giving the protein an overall molecular weight of around 57 kDa [86]. Carbohydrate moieties are attributed to 10% of the total molecular weight. Although each monomer shares interactions with the other three, avidin is commonly considered as a dimer of dimers. The dimers are constituted of monomers highly stabilized by a multitude of polar and hydrophobic interactions. The interaction between the two dimers was reported to be weaker and mainly stabilized by hydrophobic interactions. About the biological function, it has been postulated that avidin may act as a bacterial growth inhibitor in the oviduct by binding biotin helpful for bacterial growth. Also, it reported that functional avidin is found only in raw egg, as the biotin affinity for the protein is destroyed by cooking [86, 87].

5.3.1.5 Transferrins

Transferrins are a group of iron-binding glycoproteins with well-known properties found in various biological secretions [88]. Transferrins have a radius of approximately 3.6 nm in their monomeric state. Two types of transferrins are present in food matrices among the four recognized types of transferrins: (i) lactoferrin (Lf) is contained in milk and other mammalian secretions; it binds iron more tightly across a more comprehensive pH range than other transferrin proteins and has several biological activities including innate defence and (ii) ovotransferrin (Ovo) is an acidic glycoprotein that accounts for approximately 12–13% [88–90]. The iron-binding glycoprotein lactoferrin is made up of 689 amino acids and has a molar mass of about 80 kDa. It has approximately 41% α-helix and 24% β-sheets. The polypeptide chain is folded by three turns of an α-helix (residues 334–344) into two homologous lobes (N- and C-lobes) linked to each other. Each lobe consists of two domains (N1, N2 and C1, C2) forming a cleft between them; the inter-domain cleft includes a binding site for a carbonate anion-associated iron or another metallic ion- disulfide bonds stabilize the tertiary structure of Lf [88, 91]. Lf's metal-free apo form is much more versatile and more vulnerable to denaturation than the holo state, which has a very stable conformation. Lf is a simple protein with a pI of 8.6–8.9 and thus bears a positive net charge at neutral pH, allowing it to interact with acidic proteins directly. With some highly positive patches on the N-lobe and the inter-lobe area, the Lf surface's charge distribution is uneven [92]. Sialic moieties, which increase the hydration volume on the protein surface and stabilize Lf, often bear additional negative charges on the Lf surface. Lf exhibits aggregation behaviour based on ionic strength. At low ionic strength (and neutral pH), Lf molecules are mainly monomeric (radius = 3.6 nm) and positively charged. When ionic strength increased, Lf forms neutral or negatively charged nanosized aggregates of about 10–20 nm [88, 93].

Ovotransferrin is a single chain of 686 amino acids containing glycopeptides with a molecular mass of 78 kDa, a radius of 3.5 nm, and an isoelectric point of approximately 6.0. Without free sulfhydryl groups, it comprises 15 disulfide bonds. The chain is folded into two rounded lobes, N and C, like other transferrin family members, with two subdomains per lobe bound by an α-helix of nine amino acid residues that can be released through the digestion of proteases. The N- and C-lobes associate natively through noncovalent interactions, and it has been reported that isolated N- and C-lobes can reassociate spontaneously in solution. As for Lf, each lobe of Ovo can reversibly bind one Fe^{3+} ion associated with one bicarbonate anion but with different iron-binding affinities, with $K_a \sim 1.5 \times 10^{18}$ for the C-terminal lobe and 1.5×10^{14} for the N-terminal lobe [94].

Lf and Ovo display multiple biological activities in their natural forms, including antimicrobial, antiviral, antifungal, and immunomodulatory activities. Some of these operations are related to their iron-binding ability to make iron unavailable for microbial growth. Lf was also shown to be active in cell growth regulation [93, 94].

5.3.1.6 Osteopontin and Osteopontin Lactoperoxidase (OPN)

Osteopontin (OPN) is found in most tissues and body fluids as a multifunctional protein. OPN is a secreted phosphoprotein found in human milk at concentrations around 10 times higher than in bovine milk. Human OPN is a mainly unstructured glycoprotein, 36 kDa, powerfully negatively charged (pI ~3.6) [95]. In particular, phosphorylation can make post-translational changes to OPN, which raises its apparent molecular weight to around 44 kDa. OPN has been hypothesized to act as a transporter of immunomodulating and antimicrobial proteins to their place of action, given its intrinsic properties, particularly the high negative charge density [94]. Another essential protein found in milk, lactoperoxidase, is a single polypeptide chain with a molecular weight of approximately 78 kDa, a radius

Table 5.1 Example of the size on naturally formed nanoparticles between food proteins and small ligands.

Protein	Ligand	Size of formed nanoparticles (nm)	References
β-Lactoglobulin and BSA	Retinol	<5	[96]
β-Lactoglobulin	Folic acid	<10	[100]
Lactoferrin	Folic acid	16–18	[98]
β-Lactoglobulin	Lipids (oleate or linoleate)	upto 170	[97]

of 3.6 nm, and a pI of approximately 8. Eight internal disulfide bridges and about 10% of carbohydrates are found in lactoperoxidase. It plays a significant role in shielding newborns against pathogenic microorganisms in the lactating mammary gland and intestinal tract [94, 95].

5.3.2 Formation of Natural Nanostructure Subsequently to Molecular Interaction/Complexation

Proteins can form slightly higher ordered nanocomplexes through their interactions between themselves and other small molecules present in the biological fluid. Many recent studies try to generate artificially induced nanostructures with lipids, polysaccharides, proteins, or mixtures as vehicles for nutrients or bioactive substances such as minerals, vitamins, antioxidants, or fatty acids. Such complexes already exist in foods. As mentioned above, some globular proteins, for example, lactoferrin and α-lactalbumin, possess specific binding sites for mineral, iron, and calcium, suggesting a distinct physiological role. Similarly, β-lactoglobulin and BSA can complex and transport hydrophobic substances through binding specifically designed pockets with affinities ranging from 10^5 to $10^8\ M^{-1}$. Since the discovery of β-lg–vitamin A complex [96] and further extensive studies, β-lg is considered today as a natural nanovehicle forming nanocomplexes with different molecules, in particular vitamins and fatty acids [97]. More recently, we reported positively charged Lf's ability to start spontaneously nanocomplexes with negatively charged folic acid (vitamin B9) [98]. One Lf molecule bound more than 10 folic acid molecules to form complexes that self-associate into nanocomplexes with a finite size of around 16 nm. We can then assume that some proteins present in foods or food ingredients act as a naturally occurring vehicle for nutritional and bioactive substances. Interestingly, the final size of the formed nanocomplexes depends on the protein and the considered ligand (Table 5.1) but the biological functions of the nanocomplexes according to their size are still speculative. As by Le Maux et al. [99], putative roles would be (i) increase of ligand solubility and absorption, (ii) modification of the kinetics of the enzymatic hydrolysis of the protein, (iii) protection of sensitive molecules against oxidation and other stresses, and (iv) modification of the bioaccessibility of the ligands.

Nanocomplexes can also naturally occur between different proteins. The negatively charged osteopontin is highly phosphorylated in milk, enabling it to bind calcium ions and increasing its solubility consequently. Osteopontin can also form bioactive natural complexes of 5–10 nm with positively charged proteins such as lactoferrin, lactoperoxidase, and immunoglobulin [100]. Lactoferrin and osteopontin interact through a complex mechanism involving multiple cationic lactoferrin molecules binding to a single anionic molecule of osteopontin. Two distinct lactoferrin binding sites were identified, with the highest affinity constant $K_a \sim 10^6\ M^{-1}$ for the first interacting sites. Ca^{2+} binding to osteopontin or Fe^{3+} binding to lactoferrin had little effect on the overall formation of natural nanocomplexes between the two proteins. Considering that the electrostatic complementarity regions between osteopontin and lactoferrin mediate each protein's numerous biological functions, it suggested that osteopontin may act as a carrier for lactoferrin in milk and modulate the potent antimicrobial and immunostimulatory activities of the lactoferrin protein. Hence, the interaction between osteopontin and lactoferrin to form nanocomplexes could be biologically significant, depending on the individual proteins' specific biological role [94, 100].

5.3.2.1 Lipid-Based Nanostructures

Lipid Nanodroplets The lipid droplets are spherical particles formed by a triacylglycerol-rich hydrophobic core (TAG, fatty acid esters, and glycerol) covered by surface-active molecules such as proteins and phospholipids. The storage and distribution of energy require these TAG-rich droplets. The lipoproteins found in egg yolk, the nucleosomes in plant seeds, and the small fat globules in milk are examples of the natural nanostructures formed by lipids in foods. Spherical particles (15–60 nm in diameter) with a centre containing TAG and cholesterol esters surrounded by a monolayer of phospholipids and apoproteins are the lipoproteins present in egg yolk [101]. In many food products, hen egg yolk is an essential ingredient as it combines functional (emulsifying, gelling, and gelling properties), organoleptic, and nutritional properties. The natural oil droplets (0.1–10 μm in diameter) that bear structures composed of a TAG core surrounded by a monolayer of phospholipids and proteins are the nucleosomes in plant seeds. These nanostructures are subject to degradation during the ripening and germination of sources or the extraction of oil carried out by the industry [102].

Special Case: Milk Fat Globules In lipid droplets called milk fat globules, milk lipids are secreted by mammary epithelial cells [103]. Milk fat globules have a TAG core surrounded by a biological membrane formed by a phospholipid and protein trilayer. These biological assemblies consist of lipids with various chemical properties (apolar lipids and polar lipids). They are characterized by a multiscale organization of lipids that may be influenced by external factors, such as temperature. In the centre of milk fat globules, polar lipids (TAGs: fatty acid esters and glycerol; 98% of milk lipids) are found. Milk TAG molecules are rich in saturated fatty acids and are distinguished by a high melting point. Consequently, TAG milk molecules are partly crystallized below the physiological temperature (i.e. $T < 37\,°C$). Storage conditions (room temperature; fridge: 4–7 °C) cause TAG milk crystallization [104]. TAG molecules combine in the solid-state to form lamellar structures (thickness = 4–7.5 nm), which combine to form crystals and crystal networks. Polar lipids (glycerophospholipids: phosphatidylethanolamine PE, phosphatidylserine

PS, phosphatidylcholine PC, phosphatidylinositol PI, and sphingolipids, primarily sphingomyelin) along with cholesterol and proteins are found in the biological membrane covering milk fat globules. This membrane is arranged as a trilayer of polar lipids that result from the mechanisms of secretion of the milk fat globule.

In the outer bilayer of the milk fat globule membrane, a lateral arrangement of polar lipids has recently been revealed [105]. Polar lipids with a high temperature of phase transition segregate to form domains with similarities to lipid rafts in the cell membranes. The first purpose of milk fat globules is to deliver about 50% of the energy and bioactive molecules of all newborn mammals for optimum growth. Milk fat globules vary from around 50–10 μm in size (mean diameter = 4 μm). This size distribution raises concerns about the possible different biological roles of milk fat globules depending on their size.

However, since the dairy industry uses homogenization, which is a mechanical treatment capable of improving the physical stability of milk fat globules for a longer storage period of milk, milk fat globules are scarcely eaten by infants and adults as they are naturally secreted. Homogenization of milk leads to reduced milk fat globules (0.15–1 μm) and degradation of the milk protein adsorption biological membrane [106]. Regular milk fat globules and fine milk lipid droplets are structurally distinct, which may have metabolic implications [107]. Milk fat globules are also involved in manufacturing other dairy products, such as cream and cheese. Other complex lipid nanostructures corresponding to mRNA and miRNA autosomes and exosomes were found in milk [108]. There is no elucidation of their biological roles at present.

References

1 Bansal, P., Duhan, J.S., and Gahlawat, S.K. (2014). Biogenesis of nanoparticles: a review. *Afr. J. Biotechnol.* 13: 2778–2785.

2 Cho, K.H., Park, J.E., Osaka, T., and Park, S.G. (2005). The study of antimicrobial activity and preservative effects of nanosilver ingredient. *Electrochem. Acta* 51: 956–960.

3 Morones, J.R., Elechiguerra, J.L., Camacho, A. et al. (2005). The bactericidal effect of silver nanoparticles. *Nanotechnology* 16: 2346–2354.

4 Tian, J., Wong, K.K., Ho, C.M. et al. (2007). Topical delivery of silver nanoparticles promotes wound healing. *ChemMedChem* 2: 129–136.

5 Prakash, P., Gnanaprakasam, P., Emmanuel, R. et al. (2013). Green synthesis of silver nanoparticles from leaf extract of *Mimusops elengi*, Linn. For enhanced antibacterial activity against multidrug-resistant clinical isolates. *Colloids Surf., B* 108: 255–259.

6 Oves, M., Khan, M.S., Zaidi, A. et al. (2013). Antibacterial and cytotoxic efficacy of extracellular silver nanoparticles biofabricated from chromium reducing novel OS4 strain of *Stenotrophomonas maltophilia*. *PLoS One* 8: e59140. https://doi.org/10.1371/journal.pone.0059140 2013.

7 Mishra, S., Singh, B.R., Singh, A. et al. (2014). Biofabricated silver nanoparticles act as a strong fungicide against *Bipolaris sorokiniana* causing spot blotch disease in wheat. *PLoS One* 9: e97881. https://doi.org/10.1371/journal.pone.0097881.

8 Ali, S.M., Yousef, N.M.H., and Nafady, N.A. (2015). Application of biosynthesised silver nanoparticles for the control of land snail *Eobania vermiculata* and some plant

pathogenic fungi. *J. Nanomater.* 2015: 218904, 10 pages. https://doi.org/10.1155/2015/218904.

9 Hojjat, S.S. (2015). Impact of silver nanoparticles on germinated fenugreek seed. *Int. J. Agric. Crop Sci.* 8: 627–630.

10 Takkar, P.N. and Walker, C.D. (1993). The distribution and correction of zinc deficiency. In: *Zinc in Soils and Plants, Development in Plant and Soil Science* (ed. A.D. Robson), 151–165. Boston, MA: Kluwer Academic Publishers.

11 Food and Agriculture Organization of the United Nations (2000). Land, FAO and plant nutrient management, Pro Soil-Problem soil database, Rome, Italy. http://www.fao.org/ag/agl/agll/nrdb/links.jsp?lang=en (accessed 11 February 2021).

12 Alloway, B.J. (2009). Soil factors associated with zinc deficiency in crops and humans. *Environ. Geochem. Health* 31: 537–548.

13 Rashid, A. and Ryan, J. (2004). Micronutrient constraints to crop production in soils with Mediterranean-type characteristics: a review. *J. Plant Nutr.* 27: 959–975.

14 Mortvedt, J.J. (1992). Crop response to level of water-soluble zinc in granular zinc fertilisers. *Fert. Res.* 33: 249–255.

15 Gangloff, W.J., Westfall, D.G., Peterson, G.A., and Mortvedt, J.J. (2006). Mobility of organic and inorganic zinc fertilisers in soils. *Commun. Soil Sci. Plant Anal.* 37: 199–209.

16 Xie, Y., He, Y., Irwin, P.L. et al. (2011). Antibacterial activity and mechanism of action of zinc oxide nanoparticles against *Campylobacter jejuni*. *Appl. Environ. Microbiol.* 77: 2325–2331.

17 Xia, T., Kovochich, M., Brant, J. et al. (2006). Comparison of the abilities of ambient and manufactured nanoparticles to induce cellular toxicity according to an oxidative stress paradigm. *Nano Lett.* 6: 1794–1807.

18 Ryter, S.W., Kim, H.P., Hoetzel, A. et al. (2007). Mechanisms of cell death in oxidative stress. *Antioxid. Redox Signaling* 9: 49–89.

19 Long, T.C., Saleh, N., Tilton, R.D. et al. (2006). Titanium dioxide (P25) produces reactive oxygen species in immortalised brain microglia (BV2): implications for nanoparticle neurotoxicity. *Environ. Sci. Technol.* 40: 4346–4352.

20 Lovric, J., Cho, S.J., Winnik, F.M., and Maysinger, D. (2005). Unmodified cadmium telluride quantum dots induce reactive oxygen species formation leading to multiple organelle damage and cell death. *Chem. Biol.* 12: 1227–1234.

21 Lewinski, N., Colvin, V., and Drezek, R. (2008). Cytotoxicity of nanoparticles. *Small* 4: 26–49.

22 Thwala, M., Musee, N., Sikhwivhilu, L., and Wepener, V. (2013). The oxidative toxicity of Ag and ZnO nanoparticles towards the aquatic plant *Spirodela punctuta* and the role of testing media parameters. *Environ. Sci. Process. Impacts* 15: 1830–1843.

23 Panda, S.K., Chaudhury, I., and Khan, M.H. (2003). Heavy metals induce lipid peroxidation and affect antioxidants in wheat leaves. *Biol. Plant.* 46: 289–294.

24 Aarti, P.D., Tanaka, R., and Tanaka, A. (2006). Effects of oxidative stress on chlorophyll biosynthesis in cucumber (*Cucumis sativus*) cotyledons. *Physiol. Plant.* 128: 186–197.

25 Elumalai, K., Velmurugan, S., Ravi, S. et al. (2015). Green synthesis of zinc oxide nanoparticles using *Moringa oleifera* leaf extract and evaluation of its antimicrobial activity. *Spectrochim. Acta, Part A* 143: 158–164.

26 Rajiv, P., Rajeshwari, S., and Venckatesh, R. (2013). Bio-fabrication of zinc oxide nanoparticles using leaf extract of *Parthenium hysterophorus* L. and its size-dependent antifungal activity against plant fungal pathogens. *Spectrochim. Acta, Part A* 112: 384–387.

27 Bhumi, G. and Savithramma, N. (2014). Biological synthesis of zinc oxide nanoparticles from *Catharanthus roseus* (L.) G. Don. Leaf extract and validation for antibacterial activity. *Int. J. Drug Dev. Res.* 6: 208–214.

28 Sang, L., Zhao, Y., and Burda, C. (2014). TiO_2 nanoparticles as functional building blocks. *Chem. Rev.* 114: 9283–9318.

29 Owolade, O., Ogunleti, D., and Adenekan, M. (2008). Titanium dioxide affects disease development and yield of edible cowpea. *Electron. J. Environ. Agric. Food Chem.* 7: 2942–2947.

30 Khodakovskaya, M.V. and Lahiani, M.H. (2014). Nanoparticles and plants: from toxicity to activation of growth. In: *Handbook of Nanotoxicology, Nanomedicine and Stem Cell Use in Toxicology* (eds. S.C. Sahu and D.A. Casciano), 121–130. Wiley.

31 Chen, H., Seiber, J.N., and Hotze, M. (2014). ACS select on nanotechnology in food and agriculture: a perspective on implications and applications. *J. Agric. Food Chem.* 62: 1209–1212.

32 Pelaez, M., Nolan, N.T., Pillai, S.C. et al. (2012). A review on the visible light active titanium dioxide photocatalysts for environmental applications. *Appl. Catal., B* 125: 331–349.

33 Yao, K.S., Li, S.J., Tzeng, K.C. et al. (2009). Fluorescence silica nanoprobe as a biomarker for rapid detection of plant pathogens. In: *Multifunctional Materials and Structures II*, Advanced Materials Research, vol. 79–82 (eds. Y. Yin and X. Wang), 513–516. Switzerland: Trans Tech Publications.

34 Santhoshkumar, T., Rahuman, A.A., Jayaseelan, C. et al. (2014). Green synthesis of titanium dioxide nanoparticles using *Psidium guajava* extract and its antibacterial and antioxidant properties. *Asian Pac. J. Trop. Med.* 7: 968–976.

35 Rajakumar, G., Abdul Rahuman, A., Priyamvada, B. et al. (2012). *Eclipta prostrata* leaf aqueous extract mediated synthesis of titanium dioxide nanoparticles. *Mater. Lett.* 68: 115–117.

36 Chatterjee, B. and Sachdev, M. (2005). Design of a 1.7-GHz low-power delay fault-testable 32-b ALU in 180-nm CMOS technology. *IEEE Transactions on Very Large Scale Integration (VLSI) Systems,* New York, Volume 13 (11).

37 Zhang, M., Sabharwal, C.L., Tao, W. et al. (2004). Interactive DNA sequence and structure design for DNA nanoapplications. *IEEE Trans. Nanobiosci.* 3 (4): 286–292.

38 Hagiya, M. (2000). From molecular computing to molecular programming. *Proceedings of the 6th DIMACS Workshop on DNA Based Computers,* Leiden, The Netherlands, pp. 198–204.

39 Cavalcanti, A., Hogg, T., and Kretly, L.C. (2005). Transducers development for nanorobotic applications in biomedical engineering. *IEEE Conference on Nanoscale Devices and System Integration*, Houston, TX, USA.

40 Sun, J., Gao, M., and Feldmann, J. (2001). Electric field directed layer-by-layer assembly of highly fluorescent CdTe nanoparticles. *J. Nanosci. Nanotechnol.* 1 (2): 133–136.

41 Stracke, R., Böhm, K.J., Burgold, J. et al. (2000). Physical and technical parameters determining the functioning of a kinesin-based cell-free motor system. *Nanotechnology* 11 (2): 52–56.

42 Wood, W.W. (2005). Nanorobots: a new paradigm for hydrogeologic characterization? *Ground Water* 43 (4): 463.

43 Cavalcanti, A. and Freitas, R.A. Jr., (2005). Nanorobotics control design: a collective behavior approach for medicine. *IEEE Trans. Nanobiosci.* 4 (2): 133–140.

44 Boehm, F. (2006). *Nanotechnology in Environmental Applications (RNAN039A)*. BCC Research, May 2006. http://www.bccresearch.com/nano/NAN039A.asp.

45 Cavalcanti, A. (2003). Assembly automation with evolutionary nanorobots and sensor-based control applied to nanomedicine. *IEEE Trans. Nanotechnol.* 2 (2): 82–87.

46 MacNeil, J.S. (2004). Nanorobot pioneer reveal status of simulator, stem cell work. *NanoBiotech News* 2 (36): 4–5. http://www.nanorobotdesign.com/papers/nanorobotNanoBiotechNews.pdf.

47 Vasilescu, I., Kotay, K., Rus, D. et al. (2005). Data collection, storage, and retrieval with an underwater sensor network. *SenSys'05*, San Diego CA, USA.

48 Srivastava, N. and Banerjee, K. (2005). Performance analysis of carbon nanotube interconnects for VLSI applications. *IEEE/ACM ICCAD Int'l Conference On Computer-Aided Design*, pp. 383–390.

49 Bogaerts, W., Baets, R., Dumon, P. et al. (2005). Nanophotonic waveguides in silicon-on-insulator fabricated with CMOS technology. *J. Lightwave Technol.* 23 (1): 401–412.

50 Pandey, S.L., Umamageswaran, K., and Wilsey, P.A. (1999). VHDL semantics and validating transformations. *IEEE Trans. Comput. Aided Des. Integr. Circuits Syst.* 18 (7): 936–955.

51 Albert, K.J., Lewis, N.S., Schauer, C.L. et al. (2000). Cross-reactive chemical sensor arrays. *Chem. Rev.* 100 (7): 2595–2626.

52 Matzger, A.J., Lawrence, C.E., Grubbs, R.H., and Lewis, N.S. (2000). Combinatorial approaches to the synthesis of vapour detector arrays for use in an electronic nose. *J. Comb. Chem.* 2: 301–304.

53 Xu, W., Vijaykrishnan, N., Xie, Y., and Irwin, M.J. (2004). Design of a nanosensor array architecture. *ACM Proceedings of the 14th ACM Great Lakes Symposium on VLSI*, Boston, MA, USA.

54 Fung, C.K.M. and Li, W.J. (2004). Ultra-low-power polymer thin film encapsulated carbon nanotube thermal sensors. *4th IEEE Conference on Nanotechnology*, pp. 158–160.

55 Weiss, J.R.M., Menolfi, C., Morf, T. et al. (2005). Effect of body contacts on high-speed circuits in 90 nm CMOS SOI technology. *International Symposium on Signals, Circuits and Systems*. ISSCS 2005, Volume 2, pp. 537–540.

56 Bernstein, K., Chuang, C.T., Joshi, R., and Puri, R. (2003). Design and CAD challenges in sub-90nm CMOS technologies. *ICCAD '03: Proceedings of the 2003 IEEE/ACM International Conference on Computer-Aided Design*, pp. 129–136.

57 Mitra, S., Putnam, C., Gauthier, R. et al. (2005) Evaluation of ESD characteristics for 65 nm SOI technology. *2005 IEEE International SOI Conference Proceedings*, pp. 21–23.

58 Udrea, F. and Gardner, J.W. SOI CMOS gas sensors. *IEEE Transactions on Instrumentation and Measurement*, Volume 2, pp. 1379–1384.

59 Mukhopadhyay, S.C., Gooneratne, C.P., Demidenko, S., and Gupta, G.S. (2005). Low cost sensing system for dairy products quality monitoring. *2005 IEEE Instrumentationand Measurement Technology Conference Proceedings*, Ottawa, Canada, pp. 244–249.

60 Mukhopadhyay, S.C. (2004). A novel planar mesh type micro-electromagnetic sensor: Part II – Estimation of system properties. *IEEE Sens. J.* 4 (3): 308–312.

61 Nelson, C.V., Cooperman, C.B., Schneider, W. et al. (2001). Wide bandwidth time-domain electromagnetic sensor for metal target classification. *IEEE Trans. Geosci. Remote Sens.* 39 (6): 1129–1138.

62 Office of the Under Secretary of Defense for Policy (1997). Report to the Secretary of Defense on the status of Department of Defense's implementation of the United States policy on anti-personnel landmines, Washington, DC.

63 Purpura, J.W., Wynn, W.M., and Carroll, P.J. (2004). Assessment of an active electromagnetic sensor for hunting buried naval mines. *Oceans '04 MTS/IEEE Techno-Ocean '04 (IEEE Cat. No.04CH37600),* Kobe, Japan, pp. 879–889.

64 Trzaska, H. (2001). Electromagnetic field measurements in the near field. In: *Photonic EMF Measurements* (ed. H. Trzaska), 102–114. Noble Publishing.

65 Abramski, K.M. and Trzaska, H. (2002). FM EMF sensors. *IpEEE Int'l Symposium On Electromagnetic Compatibility*, Beijing, China, pp. 222–225.

66 DeHon, A. and Likharev, K.K. (2005). Hybrid CMOS/nanoelectronic digital circuits: devices, architectures, and design automation. *IEEE/ACM ICCAD Int'l Conference on Computer-Aided Design,* NW Washington, DC, USA, pp. 375–382.

67 DeHon, A. (2005). New directions for programmable devices: design of programmable interconnect for sub lithographic programmable logic arrays. *ACM Proceedings of the Int'l Symposium on Field-Programmable Gate Arrays*, Monterrey CA, USA, pp. 127–137.

68 Joannopolous, J.D., Meade, R., and Winn, J. (1995). *Photonic Crystals—Molding the Flow of Light*. Princeton, NJ: Princeton University Press.

69 Swedberg, C. (2005). N.Y. tests RFID on domestic deer. *RFID Journal.* http://www.rfidjournal.com/article/articleview/1623/1/1 (accessed 11 February 2021).

70 Swedberg, C. (2005). RFID antenna to catch fish. *RFID Journal.* http://www.rfidjournal.com/article/articleview/1966/1/1 (accessed 11 February 2021).

71 Collins, J. (2006). Taking RFID to new depths. *RFID Journal.* http://www.rfidjournal.com/magazine/article/2126 (accessed 11 February 2021).

72 Weber, W. (2003). Ambient intelligence industrial research on a visionary concept. *IEEE Int'l Symposium on Low Power Electronics and Design*, South Korea, pp. 247–251.

73 Lennartsson, R.K., Robinson, J.W.C., Persson, L. et al. (2000). Passive sonar signature estimation using bispectral techniques. *Proceedings of IEEE Workshop on Statistical Signal and Array*, Pocono Manor, PA, USA, pp. 281–285.

74 Ferguson, B.G., Lo, K.W., and Thuraisingham, R.A. (2005). Sensor position estimation and source ranging in a shallow water environment. *IEEE J. Oceanic Eng.* 30 (2): 327–337.

75 Panis, C., Hirnschrott, U., Farfeleder, S. et al. (2004). A scalable embedded DSP core for SoC applications. *IEEE Int'l Symposium on System-on-Chip,* Tampere, Finland, pp. 85–88.

76 Narendra, S.G. (2005). Challenges and design choices in nanoscale CMOS. *ACM J. Emerging Technol. Comput. Syst.* 1 (1): 7–49.

77 Horiuchi, T.K. and Cummings, R.E. (2004). A time-series novelty detection chip for sonar. *Int. J. Rob. Autom.*, ACTA Press 19 (4): 171–177. https://doi.org/10.2316/Journal.206.2004.4.206-2711.
78 Kretly, L.C., Tavora, A.A.S., and Capovilla, C.E. (2005). Design and fabrication methodology of shunt MEMS switches for wireless communication circuits. In *IMAPS Int'l Technical Symposium on Packaging, Assembling and Testing*, Campinas, Brasil.
79 Wang, A., Shi, W., Huang, J., and Yan, Y. (2016). Adaptive soft molecular self-assemblies. *Soft Matter* 12: 337–357.
80 Morris, V.J. (2010). Natural food nanostructures. In: *Nanotechnology in Food*, RSC Nanoscience & Nanotechnology, vol. 14 (eds. Q. Chaudhry, L. Castle and R. Watkins), 50–68. Cambridge: Royal Society of Chemistry.
81 Rogers, M.A. (2015). Naturally occurring nanoparticles in food. *Curr. Opin. Food Sci.* 7: 14–19.
82 Brownlow, S., Morais Cabral, J.H., Cooper, R. et al. (1997). Bovine beta-lactoglobulin at 1.8 Å resolution – still an enigmatic lipocalin. *Mol. Immunol.* 5 (4): 481–495.
83 Majorek, K.A., Porebski, P.J., Dayal, A. et al. (2012). Structural and immunologic characterisation of bovine, horse, and rabbit serum albumins. *Mol. Immunol.* 52: 174–182.
84 Permyakov, E.A. and Berliner, L.J. (2000). α-Lactalbumin: structure and function. *FEBS Lett.* 473: 269–274.
85 Huntington, J.A. and Stein, P.E. (2001). Structure and properties of ovalbumin. *J. Chromatogr. B* 756: 189–198.
86 Livnah, O., Bayer, E.A., Wilchek, M., and Sussman, J.L. (1993). Three-dimensional structures of avidin and the avidin-biotin complex. *Proc. Natl. Acad. Sci. U.S.A.* 90: 5076–5080.
87 Wu, J. and Acero-Lopez, A. (2012). Ovotransferrin: structure, bioactivities, and preparation. *Food Res. Int.* 46: 480–487.
88 Yhee, J.Y., Lee, S.J., Lee, S. et al. (2013). Tumor-targeting transferrin nanoparticles for systemic polymerised siRNA delivery in tumor-bearing mice. *Bioconjugate Chem.* 24: 1850–1860.
89 Bokkhim, H., Bansal, N., GrØndahl, L., and Bhandari, B. (2013). Physico-chemical properties of different forms of bovine lactoferrin. *Food Chem.* 141: 3007–3013. https://doi.org/10.1016/j.foodchem.2013.05.139.
90 Martins, J.T., Santos, S.F., Bourbon, A.I. et al. (2016). Lactoferrin-based nanoparticles as a vehicle for iron in food applications – development and release profile. *Food Res. Int.* 90: 16–24.
91 Sneharani, A.H., Karakkat, J.V., Singh, S.A., and Rao, A.G.A. (2010). Interaction of curcumin with β-lactoglobulin – stability, spectroscopic analysis, and molecular modeling of the complex. *J. Agric. Food Chem.* 58: 11130–11139. https://doi.org/10.1021/jf102826q.
92 Sponton, O.E., Perez, A.A., Carrara, C.R., and Santiago, L.G. (2015). Linoleic acid binding properties of ovalbumin nanoparticles. *Colloids Surf., B* 128: 219–226. https://doi.org/10.1016/j.colsurfb.2015.01.037.
93 Bouhallab, S., Lopez, C., and Axelos, M. (2017). Naturally occurring nanostructures in food. In: *Nanotechnology in agriculture and food science*, vol. 19 (eds. M.A.V. Axelos and M. Van de Voorde), 33–48. https://doi.org/10.1002/9783527697724.ch3.

94 Christensen, B. and Sørensen, E.S. (2016). Structure, function and nutritional potential of milk osteopontin. *Int. Dairy J.* 57: 1–6.

95 Futterman, S. and Hiller, J. (1972). The enhancement of fluorescence and the decreased susceptibility of enzymatic oxidation of retinol complexed with bovine serum albumin, lactoglobulin and the retinol-binding protein of human plasma. *J. Biol. Chem.* 247: 5168–5172.

96 Le Maux, S., Bouhallab, S., Giblin, L. et al. (2014). Bovine ß-lactoglobulin/fatty acid complexes: binding, structural, and biological properties. *Dairy Sci. Technol.* 94: 409–426.

97 Tavares, G.M., Croguennec, T., Lê, S. et al. (2015). Binding of folic acid induces specific self-aggregation of lactoferrin: thermodynamic characterisation. *Langmuir* 31: 12481–12488.

98 Azuma, N., Maeta, A., Fukuchi, K., and Kanno, C. (2006). A rapid method for purifying osteopontin from bovine milk and interaction between osteopontin and other milk proteins. *Int. Dairy J.* 16: 370–378.

99 Le Maux, S., Bouhallab, S., Giblin, L. et al. (2013a). Complexes between linoleate and native or aggregated beta-lactoglobulin: interaction parameters and in vitro cytotoxic effect. *Food Chem* 141 (3): 2305–2313.

100 Anton, M. (2013). Egg yolk: structures, functionalities and processes. *J. Sci. Food Agric.* 93: 2871–2880.

101 Huang, A.H.C., Qu, R., Wang, S. et al. (1987). Synthesis and degradation of lipid bodies in the scutella of maize. In: *The Metabolism, Structure, and Function of Plant Lipids* (eds. P.K. Stumpf, J.B. Mudd and W.D. Nes). Boston, MA: Springer. https://doi.org/10.1007/978-1-4684-5263-1_44.

102 Lopez, C. (2011). Milk fat globules enveloped by their biological membrane: unique colloidal assemblies with a specific composition and structure. *Curr. Opin. Colloid Interface Sci.* 16: 391–404.

103 Lopez, C., Bourgaux, C., Lesieur, P., and Ollivon, M. (2007). Coupling of time-resolved synchrotron X-ray diffraction and DSC to elucidate the crystallisation properties and polymorphism of triglycerides in milk fat globules. *Lait* 87: 459–480.

104 Lopez, C., Madec, M.N., and Jimenez, F.R. (2010). Presence of lipid rafts in the bovine milk fat globule membrane revealed by the lateral segregation of phospholipids and heterogeneous distribution of glycoproteins. *Food Chem.* 120: 22–33.

105 Lopez, C., Cauty, C., and Guyomarc'h, F. (2015). Organization of lipids in milks, infant milk formulas and various dairy products: role of technological processes and potential impacts. *Dairy Sci. Technol.* 95: 863–893.

106 Michalski, M.C. (2007). On the supposed influence of milk homogenisation on the risk of CVD, diabetes and allergy. *Br. J. Nutr.* 97: 598–610.

107 Argov-Argaman, N., Smilowitz, J.T., Bricarello, D.A. et al. (2010). Lactosomes: structural and compositional classification of unique nanometer-sized protein-lipid particles of human milk. *J. Agric. Food Chem.* 58: 11234–11242.

108 Izumi, H., Tsuda, M., Sato, Y. et al. (2015). Bovine milk exosomes contain microRNA and mRNA and are taken up by human macrophages. *J. Dairy Sci.* 98: 2920–2933.

6

Lignin Nanoparticles: Synthesis and Application

CHAPTER MENU

6.1 Overview of Lignin Nanoparticles

Lignin after cellulose is the second most abundant biopolymer on earth. According to Sameni [1] and Schutyser et al. [2], bio-refineries and pulping industries use various kraft, organosolvent, and soda enzymatic hydrolysis and steam explosion to produce lignin as a by-product. A target of 79 billion litres of second-generation biofuels needs to be made by 2022, as per the US' Energy Security and Independence Act regulated in 2007 [3]. The majority of the lignin produced today is used as boiler fuel, while a small proportion is used to manufacture value-added products [4]. The physicochemical properties of lignin are determined by the type of plant material and its processing methods. Lignin is inexpensive and has some excellent characteristics, such as high thermal stability, antioxidant activity, high carbon content, and strong stiffness [5]. However, various lignin forms have different variants in terms of functional groups, elemental composition, and molecular weight. These characteristics have increased the researcher's interest in transforming lignin into value-added products in various applications.

Nano-Technological Intervention in Agricultural Productivity, First Edition.
Javid A. Parray, Mohammad Yaseen Mir and Nowsheen Shameem.

The lignin developed from the kraft method has a size range of 10–100 μm, but the blends' mechanical properties are damaged [6]. Lignin nanoparticles (LNPs) are one of the best techniques for enhancing the mixing properties of lignin because they have new features such as elevated surface area and improved properties compared to the original material and can easily modify the surface because of the availability of a large number of functional groups such as thiols, aliphatic hydroxyl, and phenolics [7]. LNPs can play a crucial role in facilitating the valorization of lignin; nanoparticles in the polymer contribute to synthetic polymer nanoparticles.

6.2 Lignin Nanoparticle Synthesis (LNPs)

The LNPs are synthesized using a mixture of chemical and physical methods from various resources. Various methods have been discovered such as polymerization, acid precipitation, solvent exchange, ultrasonication, self-assembly, interfacial crosslinking and emulsion, microbial and enzyme-mediated, antisolvent precipitation, freeze-drying and thermal stabilization, homogenization, and alkaline precipitation, whereby lignin can be transformed into LNPs [8].

6.2.1 Polymerization

To synthesize the arabinoxylan–dehydrogenation polymer (synthetic lignin polymer) nanoparticles, coniferyl alcohol was polymerized. A mixture of coniferyl alcohol/sinapyl alcohol with two structurally related heteroxylans (HX) was used during this process. Light scattering (LS), atomic force microscopy (AFM), transmission electron microscopy (TEM), and fluorescent probes [9, 10] were used to characterize the nanoparticles produced in this process. Qian et al. [11] synthesized *N,N*-diethylaminoethyl methacrylate (DEAEMA)-grafted LNPs by atom transfer radical polymerization (ATRP) with a size range between 237 and 404 nm. The produced nanoparticles were utilized in the CO_2/N_2-switchability Pickering emulsions as a surfactant in correlation with graft density and the DEAEMA's chain length.

6.2.2 Acid Precipitation

Frangville et al. [12] developed a specific acid preparation method to manufacture novel LNPs that remain non-toxic for both yeasts and microalgae. This analysis compared two different ways for synthesizing nanoparticles that have variations in their stability upon the change in pH. In the first approach, in the presence of diluted acidic aqueous solutions, ethylene glycol solution precipitates low-sulfonated lignin. The LNPs can be retrieved like this, and their stability in various pH ranges is high. In the second process, acid-precipitated lignin is obtained from a high pH aqueous solution, which results in the formation of LNPs with stability only at the acidic level. Gupta et al. [13] also investigated the use of acid precipitation technology and synthesized the LNPs. The particle size was calculated using TEM and dynamic light scattering (DLS) and was uniformly distributed, allowing this method to be upscaled. A chemical method for the synthesis of nanoparticles using hydroxymethyl lignin was reported by Gilca et al. [14] and optimized conditions for particle size. There are many hydroxyl groups in the LNPs produced from this process, which can substitute the phenol in the synthesis of phenol–formaldehyde resin, composites, biocides, etc. Besides, the LNPs were also synthesized by the precipitation process by Richter et al. [15, 16] . Silver ions were injected into the LNPs and coated with poly(diallyldimethylammonium chloride) (PDAC). Azimvand et al. [17, 18] synthesized the LNPs using alkali lignin (AL) as a raw substrate.

6.2.3 Solvent Exchange Method

In the solvent exchange method, a solution of an organic compound in a water-miscible organic solvent, mixed with an excess of water, and nanoparticles are generated because of their decreasing solubility. This method has a wide variety of applications and is a straightforward procedure. However, the critical drawback of this method is the low level of reliable content [19]. It developed the colloidal lignin spheres of 110 nm for the first time by Qian et al. [19] based on a self-assembly method in which acetylated lignin was dissolved in tetrahydrofuran (THF) and water (67%) was gradually added to the solution. This caused the lignin molecules to associate with one another because of hydrophobic interactions. Yearla and Padmasree [20] focused on the generation of spherical dioxane lignin nanoparticles (DLNPs) and alkaline lignin nanoparticles (ALNPs) from two lignin sources such as softwood AL and hardwood dioxane lignin (DL) that was extracted from the stem region of subabul. Kai et al. [21] produced LNPs of controllable size using acetylated lignin dissolved in THF and water was added gradually to the solution.

Li et al. [22] also prepared nanocapsules in the range of 10–100 nm by dissolving the kraft lignin solution into the ethanol and adding water drop-by-drop. Because of their distinct features such as cheap preparation, simple preparation, and environmentally friendly characteristics, the nanocapsules are favoured for various applications. For the production of LNPs from waste lignin obtained from kraft pulping, Lievonen et al. [23] introduced a straightforward process. The softwood kraft lignin was dissolved in THF, and the spherical-shaped LNPs were added via dialysis after the water was introduced into the system. The nanoparticles' final size is decided based on the pre-dialysis concentration of the dissolved lignin. This study also investigated the firmness of the nanoparticle dispersion in terms of pH, salt concentration, and time. Three forms of spherical LNPs prepared by solvent exchange have been developed by Figueiredo et al. [24, 25], i.e. pure lignin nanoparticles (pLNPs), iron [III]-complexed lignin nanoparticles (Fe-LNPs), and iron oxide (Fe_3O_4) nanoparticle infused lignin nanoparticles (Fe_3O_4-LNPs), which have excellent paramagnetic behaviour and are considered the best option for cancer diagnosis and treatment. Tian et al. [26] formed a thin film by combining polyvinyl alcohol (PVA) with two forms of LNPs, namely, DLNP and ALNP, based on the technical method of lignin isolation. Sipponen et al. [27, 28] synthesized colloidal lignin particles (CLPs) by dissolving the softwood kraft lignin in THF : water ratio of 5 : 1 v/v for three hours and then precipitating by the rapid pouring of the solution into 380 g of vigorously stirred deionized water. Various weight proportions of cationic lignin were then coated onto the CLPs' insoluble fraction. In comparison, the cationic CLPs could stabilize a wide range of long-lasting Pickering emulsions than their counterparts, i.e. irregular kraft lignin particles or otherwise the regular CLPs.

Wang et al. [5] suggested a simple but unique and green technique for developing high-quantity daily LNPs. The first modified lignin was without any catalysts and solvents other than acetic anhydride through a microwave acetylation process, which served as both a reaction reagent and a dispersion solvent. Consequently, through a combination of solvent shifting and ultrasound method, high yielding standard LNPs were made. The produced LNPs can be easily separated through centrifugation and used THF can be recycled and reused, which reduce the cost of the process. Interestingly, the yield of LNPs has also been enhanced by increasing the concentration of lignin content and ultrasound strength. Xing et al. [29] produced LNPs by dissolving soda lignin in acetone and water LNPs, followed by filtration and rapidly adding filtrate in deionized water. Acetone was removed by rotary evaporation after which the solution was centrifuged to create LNPs. For the first time, the study implemented a thin layer of bioinspired melanin-like polydopamine in the LNPs that improves their biocompatibility. Zikeli et al. [30] separated the lignin from wood wastes which consisted of Iroko sawdust (IR) as well as mixed sawdust from Iroko and Norway Spruce (IRNS) and nanoparticles were produced via solvent exchange method by first dissolving lignin in dimethyl sulfoxide (DMSO) and dialysis with excess water.

6.2.4 Ultrasonication

To reduce the particle size to the nanometre scale, various mechanical treatments and methods are widely used. It does, however, have few inconveniences, including size uniformity and large particle size distributions. It is still preferable because nanoparticles are readily manufactured [31]. Gilca et al. [14] have used the sonification method to decrease the size of lignin extracted from wheat straw and sarkanda grass. The synthesized nanoparticles ranged from 10 to 50 nm and were characterized morphologically and dimensionally to detect any unusual changes in their composition and structure. This is an even more attractive approach in future applications because none of the solvents used in this process is harmful. Gonzalez et al. [32] developed the LNP through ultrasonic treatment at different times, with oscillation amplitude at 95% to achieve superior colloidal stability. Yin et al. [33] also created LNPs using the ultrasonically assisted alkaline technique by using Switcher grass as a lignin source, by dissolving it into sodium hydroxide sonication about 600 W for one hour, obtaining LNPs averaging 220 nm in length.

6.2.5 Biological Method

Lignocellulose, the most abundant renewable biomass on earth, chiefly consists of cellulose, hemicelluloses, lignin and pectin as the major constituents with resins, waxes, proteins and extractives as minor constituents. The enzyme techniques pursued in biomass degradation processes through the combinations of saccharin and fermentation processes have recently been used to increase their capacity. The improvement in material characteristics is observed when, through an enzyme-hydrolytic process, lignin is transformed into nanolignin, especially with lignocellulosic materials. For the first time, Rangan et al. [59] prepared lignin-rich cubodial nanoparticles from lignocellulosic fibers generated from Indian ridge gourd (*Luffa cylindrica*) through the breakdown of lignin-cellulose complex using specific enzymes. Juikar and Vigeshwaran [34] have also developed LNPs with cocoon fibres as a source, with the assistance of *Aspergillus* sp., a fungal isolate capable of producing high lignin peroxidase enzymes. The bulk lignin ranged from 2 to 150 and 55 μm in thickness.

The yield of nanolignin for the former was 58.4% compared with the homogenization and ultrasonic methods, whereas that was 81.4% and the latter was 64.3%. Tian et al. [35] produced LNPs of cellulolytic lignin from maize potter, poplar hardwood, and lodge poles of softwood, obtained from steam-pretreated agriculture and the overall dialysis method. The high-quality LNPs obtained were spherical, ranging from 4 to 10 with large functional groups and high pH stability, and were highly promising for applications in emerging nanomaterial fields.

6.3 Application of Lignin Nanoparticles (LNPs)

Because of their biotechnological potential in different industrial processes, LNPs have attracted considerable interest in the past decade. It will discuss some of the latest and newest applications from LNPs in this section.

6.3.1 Antibacterial Activity

Behind the antimicrobial feature of lignin, the phenolic components present in lignin play a vital role, especially the side-chain structure and functional groups. In general, the highest effectiveness against microbes is shown by double bonds in the α and β positions of the side chain with a methyl group in the γ position in the phenolic fragment. Detailed investigations were performed by Gregorová et al. [36]. On using lignin as an antimicrobial agent, the efficacy of Bjorkman lignin developed by beech wood was compared with standard additives and found the results at par. Besides, this research also explored the potential for lignin to be used as a commercial antimicrobial agent, confirming that Bjorkman lignin may theoretically be used in polyethylene films while maintaining silver's antibacterial properties. Besides, it stated that nanoscale lignin appears to be useful in treating hazardous microbes after its incorporation into silver nanoparticles. Richter et al. [15] suggested that lignin as the best choice to replace silver core since the former is abundantly available with environment-friendly characteristics. In this study, bacteria were exposed to LNPs and controlled the same for a specific period before culturing Luria–Bertani agar plates. It recorded the surprisingly outstanding performance of LNPs over controls, $AgNO_3$, and traditional AgNP, which had >12 times and 20 times excess silver ions. To find the amount of substance needed to inhibit the growth of microbes, which ultimately results in the removal of bacteria, it is essential to explore the life cycle of lignin and its degradation rate. The ability of LNPs to prevent the growth of Gram-negative bacteria such as *Erwinia carotovora* subsp. Carotovora and *Xanthomonas arboricola* PV pruni during antimicrobial tests suggested that LNPs are successful against Gram-negative bacteria compared with Gram-positive strains. These results show how this green technology can create new defence strategies against hazardous pathogens and its usefulness in the food packaging industry [37].

6.3.2 Antioxidant Activity

Owing to the complex chemical structure and functional groups in lignin, the LNPs possess significant antioxidant activity. Lu et al. [38] produced a highly concentrated solution of nanolignin, which contained higher antioxidant and free radical scavenging activities, resulting in enhanced hydrogenating capacity. Ge et al. [39] nanolignin prepared for free radical scavenging activity by the alkaline solution precipitation method shows 3.3 high fold reactivity than control due to small size, higher surface area, and bioavailability. Yearla and Padmasree [20] reported that the dioxane LNPs possess higher antioxidant properties than bulk lignin based on the data recorded for *Escherichia* coli survival rate. Domenek et al. [40] investigated the antioxidation effect in material applications. It can replace synthetic antioxidants, especially in food protection, i.e. integrating active packaging materials with natural antioxidants. The polypropylene/lignin blend antioxidant activities dependent on oxidation induction period with the help of TGA were assessed by Pouteau et al. [41]. In a polymeric matrix such as PVA, Tian et al. [26] introduced LNP to form transparent films that exhibit excellent antioxidant functions. Simultaneously, high numbers of phenolic hydroxy groups found in the LNP shell region permitted the formation of a hydrogen-bonding network that further enhanced the mechanical and thermal efficiency of the generated nanocomposite films to produce an excellent PVA matrix interfacial adhesion. Nanolignin has antioxidant properties and high dissolubility. Although highly concentrated lignin can demonstrate a significant antioxidant property, it also has some severe effects. Therefore, to reduce cytotoxicity and antioxidant activity, the concentration of lignin in biomaterials must be optimized.

6.3.3 UV Absorbents

Because of its excellent non-oxidizability capacity, lignin has tremendous sunblock potential. By the amalgamation of LNPs and pure cream, it prepared various lignin-based sunscreens. The Sun protection factor (SPF) value of pure cream is 1.03, while creams containing a different concentration of LNPs had an SPF varying from 1.26 to 2.23. With a 5% rise in LNP content, there was an improvement in SPF value to 2.23. This phenomenon has a direct relationship with the size of the LNP, i.e. the smaller the size of the LNP, the higher the sunscreen output. The probable explanation for the much higher understanding of the LNP sunscreen could be due to the lignin conjugate system that could have been generated during the nanoparticles' preparation process. Besides, simple UV-absorbing functional groups, namely, methoxyl groups and S- and G-type lignin, support this mechanism in a meaningful way. Sunscreen creams [8] also benefit from the presence of π–π stacking between the sunscreen cream and aromatic rings in LNPs. Gutiérrez-Hernández et al. [42] evaluate the photoprotection characteristics of LNPs produced from *Agave tequilana* lignin in association with ZnO nanoparticles. In the presence of a different concentration of LNPs, the SPF value was in the range of 4–13 achieved from the mixture of all the components. It observed that the company of LNPs was able to absorb both UV-B and UV-C regions that improve the SPF value of sunscreens compared to those without LNPs. The findings obtained from the UV–Vis characterization showed the better behaviour of the LNPs as a UV light barrier in grafted poly (lactic acid) (PLA), and it has also been shown that the dispersion of LNP in a PLA matrix has been encouraged by glycidyl methacrylate (GMA) grafting.

6.3.4 Hybrid Nanocomposites

LNPs in the polymer matrix and nanocomposites have also been used as reinforcing agents. Compared to original polymers, the resulting copolymers exhibit excellent mechanical and thermal properties and biocompatibility [43]. Gupta et al. [44] used LNPs as a strengthening agent for biopoly(trimethylene terephthalate)(bio-PTT) hybrid nanocomposites with 1.5% LNP weight and 7.0% carbon fibre vapour weight. Besides, an improvement in nanocomposites' mechanical characteristics, such as impact strength on thermal properties, tensile strength, tensile modulus, and biodegradation characteristics compared to bio-PTT, has been documented.

As shown by Del Saz-Orozco et al. [45], LNPs have also been used as a reinforcing agent for those materials that contain porous structures. The findings showed that because of the surfactant effect of lignin in the formulation mixture, low density and smaller cell sizes were obtained. Besides, there was a 28% increase in compressive modulus, and compared to non-reinforced foams, a 74% increase in compressive strength was observed. Qian et al. [46] used cyclohexane and AL/dioxane solution to produce lignin reverse micelle (LRM) films into high-density polyethene (HDPE). With the help of reversible addition-fragmentation chain transfer, Silmore et al. [47] formed polymer-grafted LNPs. The observed results reported that salt concentration had a major impact on zeta potential, aggregation, and interfacial stress. These causes are responsible for the alterations in both kraft lignin and polyacrylamide grafts' solubility. Besides, Chung et al. [48] modified the lignin through the reaction of graft copolymerization using lactide and in the presence of triazabicyclodecene as a catalyst. The dual function of lignin as a crosslinker to cure epoxidized natural rubber and reinforcing the filler to boost the mechanical properties of the composites of lignin/epoxidized natural rubber (ENR) was suggested by Jiang et al. [49]. They hypothesized that the presence of lignin hydroxyl and carboxyl groups could react with ENR epoxy sites via a ring-opening reaction under high-temperature dynamic heat treatment (180 °C), leading to an even dispersion of lignin along with the ENR matrix and providing excellent compatibility with rubber materials. Mahmood et al. [50] also prepared polyurethane-based lignin foam showing enhanced mechanical properties, thermal stability, curing rates, and even the flame resistance without lignin-based polyurethane foam, which suggests that the presence of lignin is advantageous in foam and rubber materials [51]. By the chemical crosslinking of acetic acid lignin with isocyanate group-terminated polyurethane monomers, Peng and Chen [52] formed pH-sensitive hydrogels whose swelling ratios improved with the increase in pH. Besides, it improved the hydrogels' thermal stability, and the ammonium sulfate release profiles indicated that the prepared hydrogels could be used as a coating. By combining hot-press processes, solvent-free radical polymerization, and microextrusion following the master batch approach, Yang et al. [37] developed LNP-reinforced poly(methyl methacrylate) (PMMA) nanocomposites.

6.3.5 Drug Delivery System

For different pharmaceutical applications such as anticancer therapy, lignin can generate nanoparticles for other substances' encapsulation [25]. Frangville et al. [12] investigated the use of non-toxic LNPs for hydrophilic compound encapsulation, i.e. rhodamine 6G with a high potential for loading. The hydrophobic compound, namely, Coumarin-6-loaded lignin microcapsule, was subsequently prepared by Tortora et al. [53] and showed that products had low cytotoxicity and could be effectively internalized into Chinese hamster ovary cells. In the presence of sodium dodecyl sulphate (SDS), there may be two concomitant explanations associated with the release of trapped Coumarin-6, such as degradation of hydrophobic lignin interactions and high molecule affinity for the new dispersion fluid. It can understand from these findings that lignin nanocapsules can be used as topical applications. If they contact lipid tissue or dermatological diseases, the pool of antioxidants and essential oils is released. Further, Chen et al. [54] developed a hydrophobic Coumarin-6-loaded pH-sensitive nanocapsule tin that can regulate the release of Coumarin-6 by altering the pH of the outer solution because of the crosslinkages of acid-labile β-thiopropionate in the capsule shell. Figueiredo et al. [24] synthesized three different forms of LNPs, namely, pLNPs, Fe-LNPs, and Fe_3O_4-LNPs with minimal cytotoxicity with varying lines of cell studied, and concerning drug loading, pLNPs were able to demonstrate their ability to effectively load poorly water-soluble drugs and other cytotoxic agents, such as sorafenib and benzazulene, and at the same time increase its release profiles in the pH range of 5.5–7.4 with sustainable manner, thereby enhancing the antiproliferation effect. Overall, the LNPs tend to be promising candidates that can be used to deliver the drug, and its Fe_3O_4-LNPs superparamagnetic activity makes it a potential and challenging player in anticancer therapy and diagnosis, such as MRI, magnetic targeting, etc. [25]. Yiamsawas et al. [55] developed a hollow nanocapsule of lignin-based polyurethane at the water–cyclohexane interface that has 2,4-toluene diisocyanate and a surfactant. Enzymatic degradation of the lignin shell may be used to achieve the release of the dye.

Interestingly, Chen et al. [56] produced LNPs using sustainable and non-toxic aqueous sodium *p*-toluenesulfonate (*p*TsONa) solution via a self-assembly process. In *p*TsONa, the amount of water-soluble or even water-insoluble drugs can be dissolved and 90% encapsulated and have sustained drug-releasing capacity. This synthesis approach with a board processing window could realize and have complete potential applications for the industrial scale-up production of LNPs.

6.3.6 Adsorbents to Remove Dyes

Azimvand et al. [17] prepared LNP-g-polyacrylic acid adsorbent in the presence of a radical initiator, namely, potassium persulfate, using copolymerization reactions between LNP and polyacrylic acid; results showed that 99 mg/g was adsorbent to LNP, while 138.88 mg/g was adsorbed to lignin nanopartum-g-polyacrylic acid. They further assessed the adsorption capacity of AL and LNPs for the removal of Basic Red 2 from aqueous solutions. Both the absorbents AL and LNPs accepted the Langmuir isotherm with values such as 55.2 and 81.9 mg/g, respectively [18].

6.3.7 As a Capacitor

The enabled lignin-based microsphere was developed by Yu et al. [57] and the lignin-based precursor sample (LAC-M and LAC-P) was started and used as a super condenser in the

6 M KOH solution. The LACs cyclic voltammetry curves at 50 mV/s showed approximately clean LAC capacitive behaviour and excellent rate efficiency as the actual current from 0.05 to 100 Ag^{-1}.

6.3.8 As a Nano-trap

The LNP–gelatin complex has shown promising results to provide a better alternative for bacterial flocculation in wastewater treatment, microbial recognition, and wastewater [33]. Xiao et al. [58] developed lignin-based nano-traps (LBNTs) for the removal of various heavy metal ions by lignin functionalization. The findings showed that LGNT achieved excellent removal efficiency in soft Hg, Cd, and borderline (Pb, Cu, and Zn ions). Besides, LGNT showed a high bactericidal rate and the antimicrobial rate LGNT with Ag towards *E. coli* and *Staphylococcus aureus*.

References

1 Sameni, J.K. (2015). Physico-chemical characterization of lignin isolated from industrial sources for advanced applications. PhD thesis. Faculty of Forestry, University of Toronto.

2 Schutyser, W., Renders, T., Van den Bosch, S. et al. (2018). Chemicals from lignin: an interplay of lignocellulose fractionation, depolymerization, and upgrading. *Chem. Soc. Rev.* 47: 852–908.

3 Agarwal, A., Rana, M., and Park, J.M. (2018). Advancement in technologies for the depolymerization of lignin. *Fuel Process. Technol.* 181: 115–132.

4 Kai, D., Tan, M.J., Chee, P.L. et al. (2016). Towards lignin-based functional materials in a sustainable world. *Green Chem.* 18: 1175–1200.

5 Wang, B., Sun, D., Wang, H.M. et al. (2019). Green and facile preparation of regular lignin nanoparticles with high yield and their natural broad-spectrum sunscreens. *ACS Sustainable Chem. Eng.* 7: 2658–2666.

6 Melro, E., Alves, L., Antunes, F.E., and Medronho, B. (2018). A brief overview on lignin dissolution. *J. Mol. Liq.* 265: 578–584.

7 Zhao, W., Simmons, B., Singh, S. et al. (2016). From lignin association to nano-/micro-particle preparation: extracting higher value of lignin. *Green Chem.* 18: 5693–5700.

8 Wang, H., Pu, Y., Ragauskas, A., and Yang, B. (2019). From lignin to valuable products – strategies, challenges, and prospects. *Bioresour. Technol.* 271: 449–461.

9 Barakat, A., Putaux, J.L., Saulnier, L. et al. (2007). Characterization of arabinoxylan-dehydrogenation polymer (synthetic lignin polymer) nanoparticles. *Biomacromolecules* 8 (4): 1236–1245.

10 Barakat, A., Gaillard, C., Lairez, D. et al. (2008). Supramolecular organization of heteroxylan-dehydrogenation polymers (syntheticlignin) nanoparticles. *Biomacromolecules* 9 (2): 487–493.

11 Qian, Y., Zhang, Q., Qiu, X., and Zhu, S. (2014). CO_2-responsive diethylaminoethyl-modified lignin nanoparticles and their application as surfactants for CO_2/N_2-switchable Pickering emulsions. *Green Chem.* 16: 4963–4968.

12 Frangville, C., Rutkevicius, M., Richter, A.P. et al. (2012). Fabrication of environmentally biodegradable lignin nanoparticles. *ChemPhysChem* 13: 4235–4243.

13 Gupta, A.K., Mohanty, S., and Nayak, S. (2014). Synthesis, characterization and application of lignin nanoparticles (LNPs). *Mater. Focus* 3: 444–454.

14 Gilca, I.A., Ghitescu, R.E., Puitel, A.C., and Popa, V.I. (2014). Preparation of lignin nanoparticles by chemical modification. *Iran. Polym. J.* 23: 355–363.

15 Richter, A.P., Brown, J.S., Bharti, B. et al. (2015). An environmentally benign antimicrobial nanoparticle based on a silver-infused lignin core. *Nat. Nanotechnol.* 10: 817–823.

16 Richter, A.P., Bharti, B., Armstrong, H.B. et al. (2016). Synthesis and characterization of biodegradable lignin nanoparticles with tunable surface properties. *Langmuir* 32: 6468–6477.

17 Azimvand, J., Didehban, K., and Mirshokraie, S.A. (2018). Safranin-O removal from aqueous solutions using lignin nanoparticle-g-polyacrylic acid adsorbent: synthesis, properties, and application. *Adsorpt. Sci. Technol.* 36 (7–8): 1422–1440.

18 Azimvand, J., Didehban, Kh., and Mirshokraie, S.A. (2018). Preparation and characterization of lignin polymeric nanoparticle using the green solvent ethylene glycol: acid precipitation technology. *BioResources* 13 (2): 2887–2897.

19 Qian, Y., Deng, Y., Qiu, X. et al. (2014). Formation of uniform colloidal spheres from lignin, a renewable resource recovered from pulping spent liquor. *Green Chem.* 16: 2156–2163.

20 Yearla, S.R. and Padmasree, K. (2016). Preparation and characterization of lignin nanoparticles: evaluation of their potential as antioxidants and U.V. protectants. *J. Exp. Nanosci.* 11: 289–302.

21 Kai, X., Chan, J., Gui-feng, L. et al. (2015). Preparation and characterization of lignin nanoparticles with controllable size by nanoprecipitation method. *Chem. Ind. For. Prod.* 5: 85–92.

22 Li, H., Deng, Y., Liu, B. et al. (2016). Preparation of nanocapsules via the self-assembly of Kraft lignin: a totally green process with renewable resources. *ACS Sustainable Chem. Eng.* 4: 1946–1953.

23 Lievonen, M., Valle-Delgado, J.J., Mattinen, M.L. et al. (2016). A simple process for lignin nanoparticle preparation. *Green Chem.* 18: 1416–1422.

24 Figueiredo, P., Lintinen, K., Kiriazis, A. et al. (2017). In vitro evaluation of biodegradable lignin-based nanoparticles for drug delivery and enhanced antiproliferation effect in cancer cells. *Biomaterials* 21: 97–108.

25 Figueiredo, P., Ferro, C., Kemell, M. et al. (2017). Functionalization of carboxylated lignin nanoparticles for targeted and pH-responsive delivery of anticancer drugs. *Nanomedicine (London)* 12 (21): 2581–2596.

26 Tian, D., Hu, J., Bao, J. et al. (2017). Lignin valorization: lignin nanoparticles as high-value bio-additive for multifunctional nanocomposites. *Biotechnol. Biofuels* 10: 192.

27 Sipponen, M.K., Smyth, M., Leskinen, T. et al. (2017). All-lignin approach to prepare cationic colloidal lignin particles: stabilization of durable Pickering emulsions. *Green Chem.* 19: 5831–5840.

28 Sipponen, M.K., Farooq, M., Koivisto, J. et al. (2018). Spatially confined lignin nanospheres for biocatalytic ester synthesis in aqueous media. *Nat. Commun.* 9: 2300.

29 Xing, Q., Buono, P., Ruch, D. et al. (2019). Biodegradable UV-blocking films through core–shell lignin–melanin nanoparticles in poly(butylenes adipate-*co*-terephthalate). *ACS Sustainable Chem. Eng.* 7: 4147–4157.

30 Zikeli, F., Vinciguerra, V., D'Annibale, A. et al. (2019). Preparation of lignin nanoparticles from wood waste for wood surface treatment. *Nanomaterials* 9: 281.

31 Beisl, S., Miltner, A., and Friedi, A. (2017). Lignin from micro- to nanosize: production methods. *Int. J. Mol. Sci.* 18: 1244–1274.

32 Gonzalez, M.N.G., Levi, M., Turri, S., and Griffini, G. (2017). Lignin nanoparticles by ultrasonication and their incorporation in waterborne polymer nanocomposites. *J. Appl. Polym. Sci.* 134: 45318.

33 Yin, H., Liu, L., Wang, X. et al. (2018). A novel flocculant prepared by lignin nanoparticles-gelatin complex from switchgrass for the capture of *Staphylococcus aureus* and *Escherichia coli*. *Colloids Surf., A* 545: 51–59.

34 Juikar, S.J. and Vigneshwaran, N. (2017). Extraction of nanolignin from coconut fibers by controlled microbial hydrolysis. *Ind. Crops Prod.* 109: 420–425.

35 Tian, D., Hu, J., Chandra, R.P. et al. (2017). Valorizing recalcitrant cellulolytic enzyme lignin via lignin nanoparticles fabrication in an integrated biorefinery. *ACS Sustainable Chem. Eng.* 5: 2702–2710.

36 Gregorová, A., Redik, S., Sedlarík, V., and Stelzer, F. (2011). Lignin-containing polyethylene films with antibacterial activity. *Proceedings in 3rd International Conference, Nanocon 2011-Conference*, TANGER Ltd. Brno, Czech Republic, EU (21–23 September 2001).

37 Yang, W., Rallini, M., Wang, D.Y. et al. (2018). Role of lignin nanoparticles in UV resistance, thermal and mechanical performance of PMMA nanocomposites prepared by a combined free-radical graft polymerization/masterbatch procedure. *Compos. Part A: Appl. Sci. Manuf.* 107: 61–69.

38 Lu, Q., Zhu, M., Zu, Y. et al. (2012). Comparative antioxidant activity of nanoscale lignin prepared by a supercritical antisolvent (SAS) process with non-nano scale lignin. *Food Chem.* 135: 63–67.

39 Ge, Y., Wei, Q., and Li, Z. (2014). Preparation and evaluation of the free radical scavenging activities of nanoscale lignin biomaterials. *BioResources* 9: 6699–6706.

40 Domenek, S., Louaifi, A., Guinault, A., and Baumberger, S. (2013). Potential of lignins as antioxidant additive in active biodegradable packaging materials. *J. Polym. Environ.* 21: 692–701.

41 Pouteau, C., Dole, P., Cathala, B. et al. (2003). Antioxidant properties of lignin in polypropylene. *Polym. Degrad. Stab.* 81: 9–18.

42 Gutiérrez-Hernández, J.M., Escalante, A., Murillo-Vázquez, R.N. et al. (2016). Use of *Agave tequilana*-lignin and zinc oxide nanoparticles for skin photoprotection. *J. Photochem. Photobiol., B* 163: 156–161.

43 Yang, M., Zhao, W., Singh, S. et al. (2019). On the solution structure of Kraft lignin in ethylene glycol and its implication for nanoparticle preparation. *Nanoscale Adv.* 1: 299–304.

44 Gupta, A.K., Mohanty, S., and Nayak, S.K. (2015). Influence of addition of vapor grown carbon fibers on mechanical, thermal and biodegradation properties of lignin nanoparticles filled bio-poly(trimethylene terephthalate) hybrid nanocomposites. *RSC Adv.* 5: 56028–56036.

45 Del Saz-Orozco, B., Oliet, M., Alonso, M. et al. (2012). Formulation optimization of unreinforced and lignin nanoparticle-reinforced phenolic foams using an analysis of variance approach. *Compos. Sci. Technol.* 72: 667–674.

46 Qian, Y., Qiu, X., Zhong, X. et al. (2015). Lignin reverse micelles for UV-absorbing and high mechanical performance thermoplastics. *Ind. Eng. Chem. Res.* 54: 12025–12030.

47 Silmore, K.S., Gupta, C., and Washburn, N.R. (2016). Tunable Pickering emulsions with polymer-grafted lignin nanoparticles (PGLNs). *J. Colloid Interface Sci.* 466: 91–100.

48 Chung, Y.L., Olsson, J.V., Lin, R.J. et al. (2013). A renewable lignin–lactide copolymer and application in biobased composites. *ACS Sustainable Chem. Eng.* 1: 1231–1238.

49 Jiang, C., He, H., Yao, X. et al. (2014). Self-crosslinkable lignin/epoxidized natural rubber composites. *J. Appl. Polym. Sci.* 131: 41166.

50 Mahmood, N., Yuan, Z., Schmidt, J., and Xu, C. (2016). Depolymerization of lignins and their applications for the preparation of polyols and rigid polyurethane foams: a review. *Renewable Sustainable Energy Rev.* 60: 317–329.

51 Hatakeyama, H., Nakayachi, A., and Hatakeyama, T. (2005). Thermal and mechanical properties of polyurethane-based geocomposites derived from lignin and molasses. *Compos. Part A: Appl. Sci. Manuf.* 36: 698–704.

52 Peng, Z. and Chen, F. (2011). Synthesis and properties of lignin-based polyurethane hydrogels. *Int. J. Polym. Mater. Polym. Biomater.* 60: 674–683.

53 Tortora, M., Cavalieri, F., Mosesso, P. et al. (2014). Ultrasound driven assembly of lignin into microcapsules for storage and delivery of hydrophobic molecules. *Biomacromolecules* 15: 1634–1643.

54 Chen, N., Dempere, L.A., and Tong, Z. (2016). Synthesis of pH-responsive lignin-based nanocapsules for controlled release of hydrophobic molecules. *ACS Sustainable Chem. Eng.* 4: 5204–5211.

55 Yiamsawas, D., Baier, G., Thines, E. et al. (2014). Biodegradable lignin nanocontainers. *RSC Adv.* 4: 11661–11663.

56 Chen, L., Zhou, X., Shi, Y. et al. (2018). Green synthesis of lignin nanoparticle in aqueous hydrotropic solution toward broadening the window for its processing and application. *Chem. Eng. J.* 346: 217–225.

57 Yu, B., Chang, Z., Zhang, Y., and Wang, C. (2018). Preparation and formation mechanism of size-controlled lignin-based microsphere by reverse phase polymerization. *Mater. Chem. Phys.* 203: 97–105.

58 Xiao, D., Ding, W., Zhang, J. et al. (2019). Fabrication of a versatile lignin-based nano-trap for heavy metal ion capture and bacterial inhibition. *Chem. Eng. J.* 358: 310–320.

59 Rangan, A., Manachiganti, M.V., Thilaividankan, R.M. et al. (2017). Novel method for the preparation of lignin-rich nanoparticles from lignocellulosic fibers. *Industrial Crops and Products* 103: 152–160.

7

Contemporary Application of Nanotechnology in Agriculture

CHAPTER MENU

7.1 Introduction

The considerable increase in crop yields, particularly for cereals, has played an essential role in meeting the world's nutritional requirements. In this regard, the increase in chemical fertilizers is a significant contributor to the rise in crop yields. The use of chemical fertilizers has multiplied fertilizer-responsive crop varieties. However, the usage of chemical fertilizers is restricted by their low efficiency of use because of the loss of fertilizer (volatilization and leaching) that contaminates the environment and raises the production costs [1]. About 50–70% of the nitrogen used by traditional fertilizers, for example, is lost to the environment [2]. Therefore, within the scientific community, alternative methods to ensure the sustainable use of nutrients are gaining considerable attention. Nanotechnology is used in this context to minimize mobile nutrient losses, grow slow-release fertilizers, and increase the accessibility of poorly accessible nutrients [3].

Nano-Technological Intervention in Agricultural Productivity, First Edition.
Javid A. Parray, Mohammad Yaseen Mir and Nowsheen Shameem.

7.2 Nanofertilizers

Nanofertilizers are nanomaterials (NMs) that are either micro- or macronutrients themselves or serve as carriers/additives (e.g. nutrients by compositing with minerals [3]). It is also possible to produce nanofertilizers by encapsulating nutrients within the NMs [2]. With higher nutrient performance, nanofertilizers increase crop yield and quality while lowering production costs, contributing to agricultural sustainability. A critical review of a nanofertilizer dataset by Kah et al. [3] showed that nanofertilizers had a median efficacy benefit of 18–29% relative to traditional fertilizers [3]. Besides, the use of phosphate nanofertilizers has been related to an improved growth rate of up to 32% and 20% of soybean seed yield compared to those treated with traditional fertilizers [4].

Plant nutrition nanotechnology's implementation encourages the production of slow/controlled release fertilizers that increase fertilizer performance, use efficiency, and reduce nutrient losses to the environment, making them environmentally friendly [5]. The effectiveness of conventional nitrogen fertilizers for fertilizers is 30–60%, whereas 80–90% of traditional phosphatic fertilizers is lost by chemical bonding in the soil and are not accessible to plants [6]. Overall, using slow-release goods decreases the amount of fertilizer. Ideally, when and where needed by plants, nanofertilizers should be able to release nutrients, limiting the conversion of extra fertilizers into gaseous forms or leaching downstream. It can be done if the release of nutrients is based on plant signalling [2], and these intelligent fertilizers can explain the contact signals between plant roots and soil microbes [6]. Plant nutrient-containing NMs react to different chemical and physical stimuli, suggesting the need for plant growth nutrients [2]. These triggers can include ethylene's development from plant roots and rhizospheric acidification released in response to soil P or K deficiency [7]. The application of nanoparticles (NPs) can alter internal root signals, as Syu et al. [7], affecting the production of ethylene from the roots of Arabidopsis. An important breakthrough towards controlled release nanofertilizers may be internal stimuli from nutrient release sources in response to P or N deficiency. Kah et al. [8] referred to the use of nano-enabled fertilizers and predicted their social acceptance, as regulatory agencies operating in many developed countries would not have any problems with nanoformulations.

Nanostructured materials are used for fertilizers' production, such as clay minerals, hydroxyapatite, chitosan, polyacrylic acid, zeolite, etc. Hydroxyapatite's large surface area and its strong interactions with urea contribute to a gradual release of N [9]. Hydroxyapatite-modified urea nanoparticles release N for up to 60 days of plant growth compared to other traditional fertilizers [9]. Ghafariyan et al. [10] used superparamagnetic iron oxide to explain an increase in chlorophyll/soybean material. A 10% increase in chlorophyll content was measured in cowpea by applying Fe nanoparticles on foliage as 0.5 g/l solution concerning the same resolution of standard forms of Fe [11]. Application of 1 and 20 mg/l ZnO-NP sprays to *Vigna radiata* [L.] and *Cicer arietinum* L. improved the plant growth compared to the conventional fertilizers [12]. Subbarao et al. [13] also reported that the K fertilizer coated with polyacrylamide polymer contributes to slow release of K. Slow-release K fertilizers are required in sandy soils to decrease the leaching and reduce the K fixation. Humic substances (HSs) stabilize the colloidal Fe by binding iron oxides [14] and inhibit the crystallization of ferric hydroxides [15]. Therefore,

Fe–HS composites have been used as nanofertilizers and are mixed with the industrial composite fertilizers containing N, P, and K [16]. Besides, compared to synthetic iron oxide nanoparticles, the use of HSs as macro-ligand provides environmental protection and increased absorption and translocation in plants. Enhanced decomposition of organic waste and compost processing can be an essential part of agricultural nanotechnology. However, the analysis is still at the beginning, and no credible result has been provided to date. It is evident from the available knowledge that nanofertilizers can minimize the number of fertilizers to be applied because of their high efficiency of usage, thus reducing the environmental impacts due to nutrient losses. Nonetheless, for safe and efficient agriculture, nanofertilizer's economic viability still needs to be studied.

7.3 Nanocomposites

The composites are made from ceramic particles (nanosize) smaller than 100 nm, which can suddenly become much more substantial. For example, grain size 10 nm metals are as much as seven times harder and more challenging than their ordinary counterparts with grain sizes in the hundreds of nanometres. A nanocomposite is a multiphase solid material where one of the phases has one, two, or three dimensions of less than 100 nm or structures having nanoscale repeat distances between the different stages that make up the material [17–19].

To improve the polymer's properties and efficiency, nanocomposites are polymers reinforced with small amounts of nanosized particles. Polymer composites with nanoclay limit the permeation of gases. Nano-zinc oxide and nano-magnesium oxide [20] are examples of polymer nanocomposites containing metal or metal oxide nanoparticles used primarily for their antimicrobial action. Cellulose nanocrystals are an attractive material to incorporate into composites because they provide highly versatile chemical functionality [21]. Nanocomposites concentrate on improving high barrier properties against oxygen diffusion, carbon dioxide, flavour compounds, and water vapour in food packaging. Bio-nanocomposites are suitable for packaging applications that include starch and cellulose derivatives [22]. Layered silicate nanoclays such as montmorillonite and kaolinite are the most promising nanoscale fillers.

In different fields, including agriculture and food processing, nanocomposites have a wide range of applications. The use of nanocomposites with new thermal and gas barrier properties will prolong food after harvest, and this application could promote food transport and storage [23]. Changes in the quantity and distribution of rainfall in dryland areas primarily affect crop production. For nanoclay composite superabsorbents, water sorption capability is an important characteristic, mainly when used under rainfed conditions. In this context, a zinc-coated nanoclay composite cross-linked polyacrylamide polymer developed to promote the rainfed rice crop has been documented for its water-holding capacity, water absorbency, and water-retaining capacity.

Compared to the initial soil, the soil's water-holding capacity with nanoclay composite was 8.5% higher [24]. Polymer nanocomposites are eco-friendly for agricultural applications. The production of mulch films in this direction may help farmers maintain moisture and control weeds [25]. For example, Nanolok™ PT is a trademark of InMat, Inc.

(Hillsborough, NJ, USA); it is a coating composed of nanodispersed silicate in polymer resin. Addition of clays and organo-modified clays into superabsorbent formulation has generated superabsorbent hydrogel nanocomposites [26]. Starch-based nanocomposite films reinforced with flax cellulose nanocrystals play an essential role in improving the mechanical and water resistance of the starch-based materials [27, 28].

Applications of biochar include carbon sequestration, soil amendment, and sorption of several classes of undesirable components from water, soil, or industrial processes [29, 30]. Biochar's nanoporous and hydrophobic properties indicate interactions with engineered nanoparticles that are being studied for accurate risk assessment of nanotoxicity [31]. Nanocomposite films of methylcellulose incorporated with nanoparticles of pediocin and zinc oxide showed antimicrobial activity against *Listeria monocytogenes* and *Staphylococcus aureus* [32].

7.4 Nanobiosensors

Biosensors are compact analytical devices consisting of biological constituents, viz., proteins, oligonucleotides, cells, and tissue, which are utilized to produce output or signals that can be used by humans. The probe, transducer, and detector are significant components of a biosensor. The investigation consists of biologically sensitized elements such as receptors, enzymes, antibodies, nucleic acids, microorganisms, and organelles, either a biologically developed or bio-mime constituent. It collects signals arising out from analyte then forwards them to the transducer. The latter measures the changes occurring because of the biological component's response and transforms the energy into calculable electrical output. Transducers are categorized as electrodes, piezoelectric crystals, optrodes, and thermistors. The signals coming out from the transducer are entrapped in the detector and then transfers to a microprocessor. After amplification, these signals are analysed, and the data collected are shifted to output to display [33]. The world's first biosensor was developed by the Research Center of Advanced Bionics of the National Institute of Advanced Industrial Science and Technology, Japan, in 2004, which can diagnose soil diseases' possible occurrence. Soil diagnosis with this biosensor was based on the principle of computing the relative activity of favourable and unfavourable soil microbes decided on the grounds of differential oxygen utilization during respiration. Biosensors incorporating nanoparticles are defined as nanobiosensors/nanosensors. The presence of NPS boosts the overall efficiency of biosensors, probably because of the increased surface for the reaction.

Nanotubes, nanowires, nanoparticles, or nanocrystals are mostly utilized to determine the signal transduction. They possess unique thermal, electrical, and optical features beneficial to increase sensitivity, decrease response period, and amend the detection range [34]. Nanosensors are replacing traditional sensors as they are portable, sensitive, and specific. Nanosensors utilizing electrochemically functional single-walled carbon nanotubes (SWCNTs) using NPs to detect gases, viz., ammonia, nitrogen oxides, sulfur dioxide, and volatile organics, have a tremendous implementation in checking agricultural pollutants [35]. Nanosmart dust and gas sensors can be used to determine the levels of environmental pollution [36]. Gas sensor arrays possessing NPs were developed through inkjet printing of metal ion chelated DNA/SWCNTs on microfabricated electrodes chased by reducing metal ions to metal. The consequences on the sensitivity and distinctively of the gas sensors towards different gases, viz., H_2, H_2S, NH_3, and NO_2, showed the increase in both parameters towards some samples by functionalizing and variable metal NPs. The united responses provide a distinct pattern for every instance that helps the system identify and quantify a particular gas [37].

7.4.1 Nanosensors in Agriculture

The essential tasks of nanotech devices include increasing the use of autonomous nanosensors placed in all cultivated fields combined with a global absolute time-positioning system and a detailed survey of crop progress that provides high-quality and critical data, such as optimum planting and crop-harvesting times, leading to better agronomic practices [38]. The sensors record temperature once each minute and are the primary moves directed to full automation of the vineyard. Crossbow Technologies has designed motes, which can be employed on the agricultural field for irrigation estimation, frost diagnosing and alert pesticide employment, harvest period, bioremediation, and water-quality computation. This could utilize nanobarcodes and nanoprocessing to keep a check on agricultural manufacturing [39]. The concept of grocery barcoding was employed to analyse and identify crop diseases efficiently and constructed nanobarcodes to tag variable pathogens in an agricultural field, which can be monitored by utilizing tools dependent on fluorescence.

7.4.2 Monitoring Soil Conditions and Plant Growth Regulators

The promotion of nanotechnology can be used for real-time monitoring of crop progression and farm status, such as moisture content, soil fertility, temperature, crop nutrient ability, pathogens, plant diseases, etc. The use of nanosensors to accurately calculate soil criteria, check pathogens, and estimate the absorption of nitrogen allows farmers to make more effective use of inputs, thus promoting sustainable agriculture [40]. Current research efforts are aimed at developing accurate water distribution systems for future irrigation. Water storage, in situ water-holding capacity, water distribution near the roots, water consumption capacity of plants, and enclosed water released according to crop requirements are necessary for this advancement. Monreal et al. [41] suggested that germs and microorganisms in wheat or other rhizosphere crops, which are an integral part of chemical signalling networks, are cross-talked. This allows microbial or plant metabolites to be released into the soil.

A brilliant nanofertilizer has been advised to possess nanobiosensors suspended in a biopolymer that coats a coating around fertilizer bits. The required nutrient release procedure depends on identifying and bonding by a nanosensor suspended in a biopolymer film covering Zn-fertilizer NPS of a specific plant SOS. The fertilizer-ZnO-NPs are released in reaction to the root signal after connection. To change the efficiency of micronutrient usage, their evaluation must be done in different crops, climate, and farm conditions after producing such nanofertilizer delivery systems, thus reducing fertilizer wastage to the environment. Nanotechnocrates possess the ability to investigate the plant's regulation of phytohormones responsible for root growth and seedling organization reacting with auxin has designed, which will help researchers understand the mechanism of plant root adaption to their environment [42]. Nanotechnology-based micro electromechanical system (MEMS) sensors can detect and react towards environmental changes by utilizing microelectronic circuits. Researchers have successfully developed wireless nanosensors consisting of MEMS cantilever beams coated with water-sensitive nanopolymer for moisture optimization.

7.4.3 Plant Pathogen Recognition

Nanochips are kinds of microarrays, consisting of fluorescent oligo capture probes used to identify hybridization. These nanochips are extremely precise and sensitive to detect the single-nucleotide changes that occur in bacteria and viruses [43]. To see the presence of *Vibrio parahaemolyticus* [43], an electrochemical immunosensor using a screen-printed electrode (SPE) with a coating of agarose/nano-Au membrane and horseradish peroxidase (HRP)-labelled VP antibody (HR-anti-VP) has been designed. A biochip sensor setup consisting of two Ti contact pads and a Ti nanowell unit on the $LiNbO_3$ substrate was developed by Seo et al. [44]. Tiny voltage changes in the nanowell showing a power spectral density were observed when the bacteria were immune from the phages (uninfected bacteria). Karnal bunt disease was recognized in the wheat crop by a nanogold-dependent immunosensor using surface plasmon resonance (SPR) [45]. Wang et al. [46] developed an electrochemical sensor to detect the pathogenic fungus *Sclerotinia sclerotiorum* by monitoring salicylic acid level in oilseeds using a modified Au-electrode with copper NPs.

7.4.4 Detection of Pesticide Residues

Nanosensors have many advantages for pesticide residue detection, such as lightweight designs, sensitivity, low detection range, super selectivity, and rapid response [46]. To verify pesticides' existence, enzyme biosensors are based on the enzyme inhibition computation used in the enzymatic reaction. Optical biosensors use multiple sensor methods to analyse biomolecular interactions, such as resonant mirrors, SPR, and waveguides. In SPR, at the surface of two media with dielectric constants with opposite signs, a surface plasmon wave occurs. Several biosensors have been developed to detect pesticide mixtures, i.e. methyl parathion [47], pyrimicarb [48], parathion [49], dichlorvos, and paraoxon [50]. Based on the inhibition of acetylcholinesterase (AChE) and chemometric calculations using artificial neural networks, certain biosensors are developed. A biosensor for organophosphorus pesticides was designed based on the multi-walled carbon nanotube (MWCNT)-modified thick film strip electrode AChE binding, which demonstrated the existence of paraoxon at parts per billion (0.145) concentrations [51]. An electrochemical sensor to detect the presence of pesticides methyl parathion and chlorpyrifos was designed, which is developed on AChE anchored on polyaniline deposited on vertically arranged carbon nanotubes (CNTs) wrapped with ssDNA. The pH around the electrode surface gets altered because of the inhibition of enzyme reaction, which helps to analyse pesticide [52]. Yu et al. [53] developed a biosensor to check the presence of organophosphorus pesticide in vegetables utilizing surface-functionalized CNTs to manage the adequate anchorage of AChE on the surface glassy carbon electrode. Tiny, fluid-filled pouches of phospholipid layers with infinite walls are liposomes. The enclosed electroactive marker immuno liposomes were used for electrochemical immunoassays as a signal enhancer. Quantum dots (QDs) are inorganic semiconductor NPs with a sphere shape with a diameter of 2–8 nm that fluoresce when a source of excitation light is stimulated. An inorganic centre that determines the colour to be released and an inorganic shell is part of its structure. Besides, there is also an aqueous organic film where biomolecules are joined to target the various biomarkers with size-dependent fluorescent properties. Therefore, there was an emission of the desired wavelengths with various configurations and amplitudes [54]. An enzyme biosensor was designed to use poly(*N*-vinyl-2-pyrrolidone) (PVP)-capped CdS QDs [49] to track trichlorfon. A colloidal Au-NP-dependent electrochemical biosensor that altered the aggregated Au-NPs on a silicate arrangement derived from sol–gel gave easy passage to electron deportation. It activated the hydrolysis reaction, increasing the amperometric response sensitivity.

An electrochemical immunosensor has also been developed to detect diuron, a substitute herbicide for phenyl urea [55]. Cheap ablated electrodes produced on a polystyrene substrate containing the Prussian blue (PB)-Au-NP coat have been modulated. In the Au-electrode's surrounding area, the electrically deposited PB-Au-NP coat improved electron transport, improving sensitivity. Kaushik et al. [56] developed a nucleic acid sensor with a coat of indium tin oxide on a chitosan nanobiocomposite coat containing iron oxide NPs accumulated on the glass surface. It has been used for pyrethroid, cypermethrin, and permethrin detection [57]. To detect the presence of metsulfuron-methyl by acquiring force curves, a nanobiosensor operating on the theory of atomic force microscopy tip operative with the acetolactate synthase enzyme was designed [58].

Besides, nanoaptamers are sensors that are synthetic single-stranded nucleic acids that can transform into distinctive and well-defined three-dimensional structures to shape binding pockets and clefts that function with a strong attraction, acting on the doctrine of a specific goal [59]. The successful aptamer sensor was developed to evaluate the toxicity status in food to detect the presence of herbicides and pesticides, respectively [60]. There have been reports of the aptamer-controlled reversible prohibition of the Au enzyme's action to see acetamiprid. For the analysis of 0.1 ppm acetamiprid, the peroxidase-like enzyme activity of Au-NPs and the intense attraction and specificity of S-18 aptamer-labelled acetamiprid were conjugated between the analysis periods of 10 minutes [61].

7.5 Nanopesticides

In commercial agriculture, the use of pesticides is a routine activity, and the production of modern, practical, and target-specific pesticides is a continuous process. Thus, a significant number of pesticides are screened annually [62]. The only minimal amount of the applied pesticides (0.1%) reaches the target pests. The remaining (99.9%), by contrast, contaminates the environment [63], which has severe implications for the food chain and human health. The pervasive presence of pesticides in the atmosphere has resulted in the production of pesticide resistance in weeds, insects, and pathogens, in addition to the effect on non-target organisms [64]. However, it is essential to remember that if there were no pesticides globally, for every life lost to pesticides, the loss of life would reach a factor of 1000 [65]. The harmful effects of synthetic pesticides tend to be minimized by bio-pesticides, but their sluggish and environmentally dependent efficacy limits their use against pests. To overcome these limitations, nanopesticides display viable potential. In the presence of suitable NMs, slow degradation and controlled release of active ingredients can provide effective pest control over a long period [66]. Therefore, for the efficient and sustainable management of various pests, nanopesticides are essential and can reduce the use of synthetic chemicals and associated environmental risks. To improve their potency, nanopesticides behave differently from traditional pesticides [8]. In dissolved and colloidal states, nanosized particles may be transported, and these types of systems have different behaviours than those of standard solutes of the same particles [67]. The solubility of active ingredients could increase the mobility and degradation of soil-inhabiting microorganisms. The solubility of AI is increased by nanoparticle-based pesticides, which are also considered to have a less harmful effect on the environment than traditional pesticides [68].

Overall, because they are applied in smaller amounts and less often than traditional pesticides, nanopesticides save energy and water. They also increase the production of pesticides and crop productivity by reducing waste and labour costs through higher yields and lower input costs. However, nanopesticides may cause health issues [69] because of different reasons described by the environmental protection agency as:

I. Dermal absorption of nanopesticides due to their size
II. Through inhalation
III. Durability and reactive potential of some NMs raise environmental concerns
IV. Scarcity of knowledge to gauge ecological exposure to engineered NMs

Different types of formulations of nanopesticides exist, and several studies have shown their efficacy against a wide variety of pests. For instance, against many sucking insects, imidacloprid (IMI) is a highly effective systemic insecticide. After 142 hours, the in vitro evaluation of IMI against *Martianus dermestid* showed 100% mortality. The nano-IMI was also more photodegradable than the traditional formulation, suggesting the environmental advantages of nano-IMI [70]. Another research showed increased use of permethrin against *Aedes aegypti* in the nanoformulation compared to the commercial form.

Moreover, the adverse effects on soil bacteria and plants in the case of nanopesticides were reduced. Similar results were observed using polyethylene glycol-based nanoformulations of carbofuran and acephate [71]. The lower toxicity of nano-acephate to non-target organisms was also regarded compared to acephate's commercial production [72] and recommended that efficiency was increased because of a slower release of the active ingredients and not because of increased uptake of nanoformulation active ingredients. Chlorfenapyr associated with silica nanoparticles exhibited two times higher insecticidal activity against cotton bollworm than chlorfenapyr associated with microparticles [73]. These results promise the slow release of active ingredients against different crop pests and can serve as useful tools in future pest management practices in agriculture. The recent studies have reported the efficacy of nano-silica to control insect pests of stored grain products [74]. Hashem et al. [75] also described the increased effectiveness and stability of *Pimpinella anisum* essential oil against red flour beetle. They established that nanoemulsions could contribute to reducing the use of harmful synthetic insecticides against insect pests.

The antimicrobial activity of nanoparticles is also well recognized against bacterial, fungal, and viral pathogens. The critical inorganic nanoparticles having pesticidal properties are silver [76], copper [77], and aluminium [78]. A study of silica–silver nanoparticles to control pathogenic fungi demonstrated that the disease-causing pathogens disappeared from the infected leaves within three days of spraying the product [79]. Antifungal silver nanoparticles were also observed against *Raffaelea* sp., causing damage to oak trees [80]. The antimicrobial activity of copper nanoparticles revealed significant antibacterial and antifungal activities [81]. The result of nanoformulated commercial fungicide (trifloxystrobin 25% + tebuconazole 50%) assessment at various concentrations against the soil-borne fungal pathogen *Macrophomina phaseolina* showed better performance in comparison with the commercial compound [82].

Weeds are emerging threats in modern agriculture. Nanoherbicides rely mostly on biodegradable polymeric substances, which could improve the efficiency of herbicides [83]. The polymeric nanoparticles encapsulated with atrazine were proven effective on *Brassica* spp. with enhanced herbicidal activity, stability, and reduced mobility in the soil compared to that of free atrazine [84]. The similar trends were observed in other studies [85] with polymer-based encapsulation of different herbicides (atrazine, ametrine, simazine, and paraquat). The increase in herbicides' bioavailability revealed similar to or slightly greater efficacy of glyphosate nanoemulsion than commercial formulation [86]. Atrazine has been known to contaminate surface and groundwater sources. The nanoformulations of atrazine and paraquat were more effective against target weeds than pure compounds. Simultaneously, genotoxicity and cytotoxicity assays revealed the less toxic impact on non-target plants such as onion [84, 85]. A study on nanoformulation of atrazine foliar interaction with Indian mustard plants revealed herbicidal activity directly through the vascular tissue of the leaves [87]. This showed great potential to maintain herbicidal activity at low concentrations and a substantial upsurge in the herbicidal efficacy. Similarly, atrazine-loaded nanocapsules were more effective against *Amaranthus viridis* L. and *Bidens pilosa* L. than commercial atrazine product [88] and thus have emerged as efficient nanoherbicides to manage the weeds as well.

The ecotoxicological life history trait data on earthworms was published by Heckmann et al. [89]. They screened different inorganic nanoparticles (Ag, Cu, Ni, Al_2O_3, SiO_2, TiO_2, and ZrO_2) for toxicity on earthworms and confirmed the toxicity of Ag, Cu, and TiO_2 nanoparticles. The total reproductive failure of earthworms was observed with both Ag treatments [89]. The earthworms can also sense and avoid the soils containing Ag nanoparticles, which suggest unpredictable effects on non-target organisms' behaviour. However, higher Ag accumulation in tissues was reported when exposed to $AgNO_3$ compared to the equivalent concentrations of Ag nanoparticles [90]. Urine et al. [91] investigated particle size's effect on the oxidation, bioavailability, and adverse effects of manufactured Cu nanoparticles in soils. They noted that oxidized Cu nanoparticles had an adverse impact in earthworms but only at relatively high concentrations. The toxicological effects of atrazine on earthworm revealed significantly different concentration-dependent LC_{50} values after additional time intervals [92]. The poly(epsilon–caprolactone) nanocapsule-based encapsulated herbicides resulted from lower toxicity to the *Pseudokirchneriella subcapitata Prochilodus lineatus* while higher toxicity to the *Daphnia similis* compared to the herbicides alone [93]. Based on risks as mentioned earlier and other nanoparticle toxicity studies [94, 95], it was evident that comprehensive risk assessment approaches for nanoparticles are needed to identify beneficial forms of nanopesticides at the cellular or molecular level.

7.6 Natural Nanoparticles: Environmental and Health Implications

7.6.1 Water Quality

Water-quality maintenance is now becoming a worldwide concern and is of crucial importance for sustainable crop production. Several industrial activities can lead to heavy metal contamination of surface and groundwater. These metals can pose a health concern for humans and animals and cause the decline of crop yield [96]. The processes for their removal from the environment have received considerable attention [97]. Several inorganic and organic adsorbents are efficient for adsorption of metal contaminants, including phyllosilicates, zeolites, fly ash, biosorbents, and activated carbon [98]. Recent studies indicate that because of their superior properties such as high surface area, adsorption ability, and unsaturated surfaces, NNPs also have potential as remediation modifications [99]. A recent study showed that Fe_3O_4 NPs could be useful in removing aqueous solution contaminants [100].

Arsenic contamination is a global problem in drinking water, occurring predominantly as As^{3+}(H_3AsO_3, H_2AsO^{3-}, and $HAsO_3{}^{2-}$) and As(V)(H_3AsO_4, H_2AsO^{4-}, and $HAsO_3{}^{2-}$) [101]. Activated C and oxides of Mn, Fe, and Zr are naturally accessible nanoscale adsorbents that have been commonly used for As remediation. Manganese oxide can also be used to remove As^{3+} from the atmosphere as an oxidizing agent [102]. For drinking water elimination, a combination of MnO_2 and Fe_3O_4 for oxidation, adsorption, and magnetic reduction can be effective [103].

The significance of Fe^{2+} and Fe^{3+} NPs in the adsorption of As and their role in the microbial cycling of As and Fe [104] has been recorded in a recent report. Significant attention has been given to the use of magnetically assisted chemical speciation of NPS iron oxides for the decontamination process in water treatment [105]. In the presence of competing anions such as phosphate [106], adsorption of As and Cr by mixed magnetite and maghemite NPs from aqueous solution is a promising way to remediate groundwater. The fundamental principle is that electrostatic attraction between heavy metals and the magnetite mixture depends on pH, contact time, and initial concentration of dissolved and adsorbent species [106]. As a nanoadsorbent for removing Cd from drinking water, iron oxide NNPs are efficient [107]. Several NNPs for removing metal ions have been tested to boost water quality, such as maghemite NPs [108]. Iron oxide NNPs have been functionalized and stabilized with humic acid [109] and amino-functionalized polyacrylic acid [110]. Non-mobile soil nano-NPS effectively minimize the transport of labile P and thus increase the quality of water [111]. NPS can serve as vectors for transporting nutrients from the soil to surface waterways [112]. By strongly sorbing them and subsequently acting as mobile vectors, these colloidal particles ease the transport of trace metal contaminants [113]. Pollutants, including cationic, anionic, nonpolar organic, and polar-organic compounds, can be present in several forms in the atmosphere.

On the other hand, colloidal particles are mainly minerals with amphoteric surfaces, fixed charge surfaces, and mineral phases covered with organic matter [114]. Surface complexation, ion exchange, and hydrophobic partitioning [114] are the essential sorption mechanisms involved in the colloid-facilitated pollutants' transport. Nanoscale particles may exercise important trace metal transport control either by retarding the vehicle when the particles are static in the matrix or by encouraging transport when they are mobile [113]. For remediation by degrading or immobilizing redox-sensitive pollutants, biocolloid transport is also essential [115]. Efficiency in extracting hydrophobic contaminants from sediments was also demonstrated by natural organic colloids, such as HSs [116]. The dominant factor regulating bioavailability and the conversion of pollutants may be NPs [117]. The main processes regulating the filtration of contaminants, especially organic ones, are adsorption and desorption. By altering the chemical speciation of metal ions, i.e. Cr, Pu, Co, Pb, Cu, As, and Se [118], redox reactions are also responsible for filtration contaminants from the soil water systems. These reactions also account for the conversion and reactivity of Fe^{-} and Mn oxides which, in natural nanorange systems such as HSs, have an enormous capacity to absorb metal pollutants and various dissolved organic components [119].

7.6.2 Interactions with Contaminants and Other Organisms

Figure 7.1 indicates the interactions of NPs with different pollutants and species. In sediments, natural organic colloids constitute around 9% of the total organic C and have a high sorption potential for organic compounds [120]. Restricted potential for microbial degradation has been observed, and up to two orders of magnitude can decrease uptake in organisms [121]. NPS can interact with an array of minerals in the soil in different ways. The degree of reactions that regulate the fate of nutrients and pollutants can also differ as a result. NPS

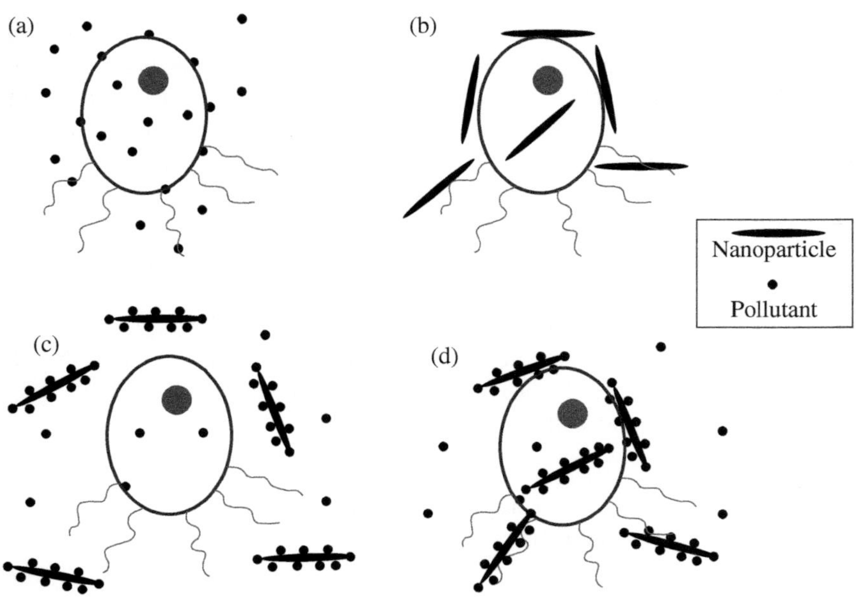

Figure 7.1 Schematic representation of the interaction of NPs with pollutants and organisms. (a) Adsorption and uptake of pollutant, (b) adsorption and uptake of nanoparticle, (c) adsorption (or absorption) of pollutants onto NP and reduction in pollutant uptake by organisms, and (d) adsorption of NP with adsorbed pollutant and possible uptake of pollutant-NP. Source: Nowack and Bucheli [120]. © 2007, Elsevier.

interactions with nutrients and contaminants and the environment rely on the growth, stability, phase transition, aggregation, and ageing of nanoparticles [122].

7.6.3 Environmental Risks and Biogeochemistry of NNPs

NPS [123] environmental risk assessment remains subject to various problems, including chemical characterization of test products, reference NMs for toxicity control, improvements in test methodology, additional measurements within existing experiments, and quantification of novel or unusual toxicological characteristics. Challenges exist in categorizing NPS concerning their impact on life cycles and deciding their entry points into the environment [124]. Although the chemical characterization of NPs is difficult, in regulatory toxicity testing, requirements for the classification of these particles can be set. The analytical capacity to describe NPS regularly for ecotoxicology [125] has apparent information gaps. NPS can reach the body through various routes and are widely distributed to different biological systems' organs and tissues. The most crucial exposure mode to nano-NPS is inhalation. Essential exposure channels [126] are both ingestion and skin transmission. One of the major sources of NPs that can affect biological systems is wastewater. Traces of pharmaceuticals associated with NPs can be entrained in wastewater and subsequently enter the environment. Other sources of NPs include dust particles from soil, construction sites, and industrial activities [127]. The natural system can tolerate a background concentration of 5000–10 000 airborne NPs/cm^3, whereas, at pollution levels, this number can exceed 3 000 000 NPs/cm^{-3}.

Insoluble airborne iridium NPs of the size range of 15–80 nm can be deposited in the gastrointestinal region. It has been demonstrated that gene expression profiles caused by pulmonary injury and inflammation were related to the increased expressions and subsequent contributions from cytokines and chemokines because of the exposure to urban particulate matter [128]. NPS generally store in specific locations such as mitochondria and vesicles, where they can generate reactive oxygen species and be toxic [129]. Inflammation and fibrosis are the main effects found at the organism's level, while oxidative stress, antioxidant activity, and cytotoxicity are cellular effects [130]. Hydroponic experiments using aluminium oxide have also reported root elongation [131]. Cellular internalization of nanoscale or micro-scale particles, including endocytosis and phagocytosis [132], is processed at the cellular level by eukaryotes such as protist metazoans. To localize the NPS into the cell mitochondria, unicellular protozoa have also been reported [133].

Ecological health can potentially affect NPs in three ways:

i. Direct impact on species
ii. Impact on environmental environments that are non-living.
iii. Contaminant and nutrient interactions.

7.6.4 Environmental Issues

NPS plays an essential role in creating dust clouds by coagulating from various sources after their release into the atmosphere. A recent study in South Asia found that 70% of brown clouds are particulate matter derived from biomass combustion, consisting primarily of C-NPs derived from unprocessed fuels such as cooking wood and animal dung [134]. This brown cloud leads to glacial melting, decreased sunshine, severe weather conditions affecting agricultural production, and a reduction in the Indian monsoon season [135]. The OH-radicals, which are known to be reactive free radicals in the photochemical degradation of organic contaminants in the atmosphere, are theoretically decreased because of nano-NPS binding [136]. Nano-NPS is also inadvertently released from hydrogen fuel cells and molecular hydrogen, going up to the stratosphere and creating an excess of water. This water can cause stratosphere cooling, encourage heterogeneous ozone-destroying chemistry, increase noctilucent clouds, and changes in tropospheric chemistry and atmospheric–biosphere interaction [137].

7.7 Future Perspective

Sustainable agriculture, where abiotic–biotic-living beings live in compliance with the traditional stability of food chains and their associated energy balances, must be taken as an ecosystem strategy. To improve productivity in agriculture, emerging technology, modernization, increased use of nanochemicals, specialization, and government policies are being adapted. Therefore, establishing the latest technology in the food industry is mandatory. Therefore, new and future technology is nanotechnology that possesses unique property in the food supply chain around the world agricultural sector. Some focused areas may need more attention in the near future research studies in the field of agrarian nanotechnology or nanofoods:

1. New environmental and protection distribution methods for the transport of specific food compounds, nutrients from plants, etc.
2. Biosensor-related nanotechnology plays an essential role in the management of insect pests and agricultural food products.
3. To detect, validate, and access the effect of each NM/nanofood in entire ecosystems, new analytical methods are required to be developed. NMs or nanofoods should perform life cycle research.

References

1 FAO (2017). *The Future of Food and Agriculture "Trends and Challenges"*. Food and Agriculture Organization of the United Nations Rome.
2 DeRosa, M.C. (2010). Nanotechnology in fertilizers. *Nat. Nanotechnol.* 5: 91.
3 Kah, M., Kolkata, R.S., Gogos, A., and Bucheli, T.D. (2018). A critical evaluation of nano pesticides and nano fertilizer against their conventional analogues. *Nat. Nanotechnol.* 13: 677–684.
4 Liu, R. and Lal, R. (2014). Synthetic apatite nanoparticles as a phosphorus fertilizer for soybean (*Glycine max*). *Sci. Rep.* 4: 5686.
5 Rico, C.M., Majumdar, S., Duarte-Gardea, M. et al. (2011). Interaction of nanoparticles with edible plants and their possible implications in the food chain. *J. Agric. Food Chem.* 59: 3485–3498.

6 Rop, K., Karuku, G.N., Mbui, D. et al. (2019). Evaluating the effects of formulated nano-NPK slow-release fertilizer composite on the performance and yield of maize, kale and capsicum. *Ann. Agric. Sci.* 64: 9–19.

7 Syu, Y.Y., Hung, J.H., Chen, J.C., and Chuang, H.W. (2014). Impacts of size and shape of silver nanoparticles on *Arabidopsis* plant growth and gene expression. *Plant Physiol. Biochem.* 83: 57–64.

8 Kah, M., Tufenkji, N., and White, J.C. (2019). Nano-enabled strategies to enhance crop nutrition and protection. *Nat. Nanotechnol.* 14: 532–540.

9 Kottegoda, N., Munaweera, I., Madusanka, N., and Karunaratne, V. (2011). A green slow-release fertilizer composition based on urea-modified hydroxyapatite nanoparticles encapsulated wood. *Curr. Sci.* 101: 73–78.

10 Ghafariyan, M.H., Malakouti, M.J., Dadpour, M.R. et al. (2013). Effects of magnetite nanoparticles on soybean chlorophyll. *Environ. Sci. Technol.* 47: 10645–10652.

11 Karami, M., Bahabadi, M.A., Delfani, S., and Ghozatloo, A. (2014). A new application of carbon nanotubes nanofluid as working fluid of low-temperature direct absorption solar collector. *Sol. Energy Mater. Sol. Cells* 121: 114–111.

12 Mahajan, P., Dhoke, S., and Khanna, A. (2011). Effect of nano-ZnO particle suspension on growth of mung (*Vigna radiata*) and gram (*Cicer arietinum*) seedlings using plant agar method. *J. Nanotechnol.* 2011: 54–61.

13 Subbarao, C.V., Kartheek, G., and Sirisha, D. (2013). Slow release of potash fertilizer through polymer coating. *Int. J. Appl. Sci. Eng.* 11: 25–30.

14 Angelico, R., Ceglie, A., He, J.Z. et al. (2014). Particle size, charge and colloidal stability of humic acids coprecipitated with ferrihydrite. *Chemosphere* 99: 239–247.

15 Usman, M., Byrne, J.M., Chaudhary, A. et al. (2018). Magnetite and green rust: synthesis, properties, and environmental applications of mixed-valent iron minerals. *Chem. Rev.* 118: 3251–3304.

16 Sorkina, T.A., Polyakov, A.Y., Kulikova, N.A. et al. (2014). Nature-inspired soluble iron-rich humic compounds: new look at the structure and properties. *J. Soils Sediments* 14: 261–268.

17 Ajayan, P.M., Schadler, L.S., and Braun, P.V. (eds.) (2003). *Nano-Composite Science and Technology*. Weinheim: Wiley-VCH.

18 Kumar, S.K. and Krishnamoorti, R. (2010). Nanocomposites: structure, phase behavior, and properties. *Annu. Rev. Chem. Biomol. Eng.* 1: 37–58.

19 Ke, Y.C. and Stroeve, P. (2005). *Polymer-Layered Silicate and Silica Nanocomposites*, 1e. Amsterdam: Elsevier BV.

20 Chaudhry, Q., Castle, L., and Watkins, R. (eds.) (2010). *Nanotechnologies in Food*. Cambridge: Royal Society of Chemistry.

21 Cao, X., Habibi, Y., Magalhães, W.L.E. et al. (2011). Cellulose nanocrystals-based nanocomposites: fruits of a novel biomass research and teaching platform. *Curr. Sci.* 100 (8): 1172–1176.

22 Tiwari, A. (2010). *Recent Developments in Bio-Nanocomposites for Biomedical Applications*, 1e. Hauppauge, NY: Nova Science Publishers, Inc.

23 Sozer, N. and Kokini, J.L. (2009). Nanotechnology and its applications in the food sector. *Trends Biotechnol.* 27 (2): 82–89.

24 Cortes-Lobos, R. (2013). Nanotechnology research in the US agri-food sectoral system of innovation: towards sustainable development. dissertation. Atlanta, GA: Georgia Institute of Technology. https://smartech.gatech.edu/bitstream/handle/1853/47541/cortes_rodrigo_f_201305_phd.pdf?sequence=1&isAllowed=y

25 Jatav, G.K., Mukhopadhyay, R., and De, N. (2013). Characterization of swelling behaviour of nanoclay composite. *Int. J. Innov. Res. Sci. Eng. Technol.* 2 (5): 1560–1563.

26 Ray, S.S. (2013). *Environmentally Friendly Polymer Nanocomposites: Types, Processing and Properties.* Cambridge: Woodhead Publishing.

27 Kabiri, K., Omidian, H., Zohuriaan-Mehr, M.J., and Doroudiani, S. (2011). Superabsorbent hydrogel composites and nanocomposites: a review. *Polym. Compos.* 32 (2): 277–289.

28 Cao, X.D., Chen, Y., Chang, P.R. et al. (2008). Starch-based nanocomposites reinforced with flax cellulose nanocrystals. *Express Polym. Lett.* 2 (7): 502–510.

29 Peterson, S.C., Jackson, M.A., and Appell, M. (2013). Biochar: sustainable and versatile. In: *Advances in Applied Nanotechnology for Agriculture* (eds. B. Park and M. Appell), 193–205. Washington, DC: American Chemical Society.

30 Joseph, S., Graber, E.R., Chia, C. et al. (2013). Shifting paradigms: development of high-efficiency biochar fertilizers based on nanostructures and soluble components. *Carbon Manag.* 4 (3): 323–343.

31 Pignatello, J.J., White, J.C., and Uchimiya, M. (2013). Nanoscale interactions between engineered nanomaterials and black carbon [biochar] in soil. *NSF Nanoscale Science and Engineering Grantees Conference*, Arlington, VA.

32 Espitia, P.J., Soares, N.F., Teófilo, R.F. et al. (2013). Physical-mechanical and antimicrobial properties of nanocomposite films with pediocin and ZnO nanoparticles. *Carbohydr. Polym.* 94 (1): 199–208.

33 Velasco-Garcia, M.N. (2014). Optical biosensors for probing at the cellular level: a review of recent progress and future prospects. *Semin. Cell Dev. Biol.* 20 (1): 27–33.

34 Yao, J., Yang, M., and Duan, Y.X. (2014). Chemistry, biology, and medicine of fluorescent nanomaterials and related systems: new insights into biosensing, bioimaging, genomics, diagnostics, and therapy. *Chem. Rev.* 114: 6130–6178.

35 Wanekaya, A.K., Chen, W., Myung, N.V., and Mulchandani, A. (2006). Nanowire-based electrochemical biosensors. *Electroanalysis* 18 (6): 533–550.

36 Mousavi, S.R. and Rezaei, M. (2010). Nanotechnology in agriculture and food production. *J. Appl. Environ. Biol. Sci.* 1 (10): 414–419.

37 Su, H.C., Zhang, M., Bosze, W. et al. (2013). Metal nanoparticles and DNA co-functionalized single-walled carbon nanotube gas sensors. *Nanotechnology* 24 (50): 502–505.

38 El Beyrouthya, M. and El Azzi, D. (2014). Nanotechnologies: novel solutions for sustainable agriculture. *Adv. Crop Sci. Technol.* 2: e118.

39 Li, Y., Cu, Y.T., and Luo, D. (2005). Multiplexed detection of pathogen DNA with DNA-based fluorescence nanobarcodes. *Nat. Biotechnol.* 23 (7): 885–889.

40 Prasad, R., Pandey, R., and Barman, I. (2016). Engineering tailored nanoparticles with microbes: quo Vadis. *WIREs Nanomed. Nanobiotechnol.* 8: 316–330.

41 Monreal, C.M., DeRosa, M., Mallubhotla, S.C. et al. (2015). The Application of Nanotechnology for Micronutrients in Soil-Plant Systems. *VFRC Report 2015/3*. Washington, DC: Virtual Fertilizer Research Center, p. 44.

42 McLamore, E.S., Diggs, A., Calvo, M.P. et al. (2010). Non-invasive quantification of endogenous root auxin transport using an integrated flux microsensor technique. *Plant J.* 63 (6): 1004–1016.

43 Lopez, M.M., Llop, P., Olmos, A. et al. (2009). Are molecular tool solving the challenges posed by detection of plant pathogenic bacteria and viruses? *Curr. Issues Mol. Biol.* 11: 13–46.

44 Seo, S., Dobozi-King, M., Young, R.F. et al. (2008). Patterning a nanowell sensor bio-chip for specific and rapid detection of bacteria. *Microelectron. Eng.* 85 (7): 1484–1489.

45 Singh, S., Singh, M., Taj, G. et al. (2012). Development of surface plasmon resonance (SPR) based immuno-sensing system for detection of fungal teliospores of Karnal bunt (*T. indica*), a quarantined disease of wheat. *J. Biosens. Bioelectron.* 3: 1–8.

46 Wang, Z., Wei, F., Liu, S.Y. et al. (2010). Electrocatalytic oxidation of phytohormone salicylic acid at copper nanoparticles-modified gold electrode and its detection in oilseed rape infected with fungal pathogen *Sclerotinia sclerotiorum*. *Talanta* 80: 1277–1281.

47 Kang, T.F., Wang, F., Lu, L.P. et al. (2010). Methyl parathion sensors based on gold nanoparticles and nafion film modified glassy carbon electrodes. *Sens. Actuators, B* 145: 104–109.

48 Sun, H. and Fung, Y. (2006). Piezoelectric quartz crystal sensor for rapid analysis of pirimicarb residues using molecularly imprinted polymers as recognition elements. *Anal. Chim. Acta* 576: 67–76.

49 Li, C., Wang, C., and Hua, S. (2006). Development of a parathion sensor based on molecularly imprinted nano-TiO_2 self-assembled film electrode. *Sens. Actuators, B* 117: 166–171.

50 Tuoro, D., Portaccio, M., Lepore, M. et al. (2011). An acetylcholinesterase biosensor for determination of low concentrations of paraoxon and dichlorvos. *New Biotechnol.* 29: 132–138. https://doi.org/10.1016/j.nbt.2011.04.011.

51 Joshi, K.A., Tang, J., Haddon, R. et al. (2005). A disposable biosensor for organophosphorus nerve agents based on carbon nanotubes modified thick film strip electrode. *Electroanalysis* 17: 54–58.

52 Viswanathan, S., Radecka, H., and Radecki, J. (2009). Electrochemical biosensor for pesticides based on acetylcholinesterase immobilized on polyaniline deposited on vertically assembled carbon nanotubes wrapped with ssDNA. *Biosens. Bioelectron.* 24: 2772–2777.

53 Yu, Y., Anthony, S.M., Bae, S.C. et al. (2009). Biomolecular science of liposome nanoparticle constructs. *Mol. Cryst. Liq. Cryst.* 507: 18–25.

54 Patolsky, F., Zheng, G., and Lieber, C. (2006). Nanowire-based biosensors. *Anal. Chem.* 78: 4260–4269.

55 Sharma, P., Sablok, K., Bhalla, V., and Suri, C.R. (2011). A novel disposable electrochemical immunosensor for phenyl urea herbicide Diuron. *Biosens. Bioelectron.* 26 (10): 4209–4212.

56 Kaushik, A., Solanki, P.R., Ansarib, A.A. et al. (2009). Iron oxide-chitosan hybrid nanobiocomposite based nucleic acid sensor for pyrethroid detection. *Biochem. Eng. J.* 46: 132–140.
57 Ramanathan, M., Luckarift, H.R., Sarsenova, A. et al. (2009). Lysozyme-mediated formation of protein–silica nanocomposites for biosensing applications. *Colloids Surf., B* 73: 58–64.
58 da Silva, A.C., Deda, D.K., da Roz, A.L. et al. (2013). Nanobiosensors based on chemically modified AFM probes: a useful tool for metsulfuron-methyl detection. *Sensors (Basel)* 13 (2): 1477–1489.
59 Chiu, T.C. and Huang, C.C. (2009). Aptamer-functionalized nano-bio sensors. *Sensors* 9: 10356–10388.
60 McKeague, M., Giamberardino, A., and De Rosa, C.M. (2011). Advances in aptamer-based biosensors for food safety. In: *Environmental Biosensors* (ed. V. Somerset). Janeza Trdine 9, 51000 Rijeka, Croatia– EUROPEAN UNION: INTECH. ISBN: 978-953-307-4863.
61 Weerathunge, P., Ramanathan, R., Shukla, R. et al. (2014). Aptamer-controlled reversible inhibition of gold nanozyme activity for pesticide sensing. *Anal. Chem.* 86 (24): 11937–11941.
62 Resh, V.H. and Cardé, R.T. (2009). *Encyclopedia of Insects.* Academic Press.
63 Carriger, J.F., Rand, G.M., Gardinali, P.R. et al. (2006). Pesticides of potential ecological concern in sediment from South Florida canals: an environmental risk prioritization for aquatic arthropods. *Soil Sediment Contam.* 15: 21–45.
64 Rai, M. and Ingle, A. (2012). Role of nanotechnology in agriculture with special reference to management of insect pests. *Appl. Microbiol. Biotechnol.* 94: 287–293.
65 Lomborg, B. (2002). *The Skeptical Environmentalist. 2001*, 515. Cambridge: Cambridge University Press.
66 Chhipa, H. (2017). Nano fertilizers and nano pesticides for agriculture. *Environ. Chem. Lett.* 15: 15–22.
67 Kumpiene, J., Lagerkvist, A., and Maurice, C. (2008). Stabilization of As, Cr, Cu, Pb and Zn in soil using amendments–a review. *Waste Manag.* 28: 215–225.
68 Kah, M. and Hofmann, T. (2014). Nanopesticide research: current trends and future priorities. *Environ. Int.* 63: 224–235.
69 Ragaei, M. and Sabry, A.-k. (2014). Nanotechnology for insect pest control. *Int. J. Sci. Environ. Technol.* 3: 528–545.
70 Guan, Y., Pearce, R.C., Melechko, A.V. et al. (2008). Pulsed laser dewetting of nickel catalyst for carbon nanofiber growth. *Nanotechnology* 19: 235604.
71 Pradhan, S., Roy, I., Lodh, G. et al. (2013). Entomotoxicity and biosafety assessment of PEGylated acephate nanoparticles: a biologically safe alternative to neurotoxic pesticides. *J. Environ. Sci. Health B* 48: 559–569.
72 Pankaj, Shakil, N.A., Kumar, J. et al. (2012). Bioefficacy evaluation of controlled release formulations based on amphiphilic nanopolymernanopolymer of carbofuran against Meloidogyne incognita infecting tomato. *J. Environ. Sci. Health B* 47: 520–528.
73 Song, M.R., Cui, S.M., Gao, F. et al. (2012). Dispersible silica nanoparticles as carrier for enhanced bioactivity of chlorfenapyr. *J. Pest. Sci.* 37: 258–260.

74 Gamal, Z.M.M. (2018). Nanoparticles: a recent approach for controlling stored grain insect pests. *Acad. J. Agric. Res.* 6: 088–094.
75 Hashem, A.S., Awadalla, S.S., Zayed, G.M. et al. (2018). *Pimpinella anisum* essential oil nanoemulsions against *Tribolium castaneum*—insecticidal activity and mode of action. *Environ. Sci. Pollut. Res.* 25: 18802–18812.
76 Kim, H.J., Phenrat, T., Tilton, R.D., and Lowry, G.V. (2012). Effect of kaolinite, silica fines and pH on transport of polymer-modified zero-valent iron nanoparticles in heterogeneous porous media. *J. Colloid Interface Sci.* 370: 1–10.
77 Gogos, A., Knauer, K., and Bucheli, T.D. (2012). Nanomaterials in plant protection and fertilization: current state, foreseen applications, and research priorities. *J. Agric. Food Chem.* 60: 9781–9792.
78 Stadler, T., Buteler, M., Weaver, D.K., and Sofie, S. (2012). Comparative toxicity of nanostructured alumina and a commercial inert dust for *Sitophilus oryzae* (L.) and *Rhyzopertha Dominica* (F.) at varying ambient humidity levels. *J. Stored Prod. Res.* 48: 81–90.
79 Park, H.J., Kim, S.H., Kim, H.J., and Choi, S.H. (2006). A new composition of nanosized silicasilver for control of various plant diseases. *Plant Pathol. J.* 22: 295–302.
80 Kim, S.W., Kim, K.S., Lamsal, K. et al. (2009). An in vitro study of the antifungal effect of silver nanoparticles on oak wilt pathogen *Raffaelea* sp. *J. Microbiol. Biotechnol.* 19: 760–764.
81 Esteban-Tejeda, L., Malpartida, F., Esteban-Cubillo, A. et al. (2009). Antibacterial and antifungal activity of a soda-lime glass containing copper nanoparticles. *Nanotechnology* 20: 505701.
82 Kumar, G.D., Natarajan, N., and Nakkeeran, S. (2016). Antifungal activity of nanofungicide trifloxystrobin 25%+ tebuconazole 50% against *Macrophomina phaseolina*. *Afr. J. Microbiol. Res.* 10: 100–105.
83 Abigail, E.A. and Chidambaram, R. (2017). Nanotechnology in herbicide resistance. *Nanostruct. Mater.* 207: 207–212.
84 Pereira, A.E., Grillo, R., Mello, N.F. et al. (2014). Application of poly(epsilon-caprolactone) nanoparticles containing atrazine herbicide as an alternative technique to control weeds and reduce damage to the environment. *J. Hazard. Mater.* 268: 207–215.
85 Grillo, R., Pereira, A.E., Nishisaka, C.S. et al. (2014). Chitosan/tripolyphosphate nanoparticles loaded with paraquat herbicide: an environmentally safer alternative for weed control. *J. Hazard. Mater.* 278: 163–171.
86 Lim, C.J., Basri, M., Omar, D. et al. (2013). Green nanoemulsion-laden glyphosate isopropylamine formulation in suppressing creeping foxglove (*A. gangetica*), slender buttonweed (*D. ocimifolia*) and buffalo grass (*P. conjugatum*). *Pest Manag. Sci.* 69: 104–111.
87 Bombo, A.B., Pereira, A.E.S., Lusa, M.G. et al. (2019). A mechanistic view of interactions of a nanoherbicide with target organism. *J. Agric. Food Chem.* 67: 4453–4462.
88 Sousa, G.F., Gomes, D.G., Campos, E.V. et al. (2018). Post-emergence herbicidal activity of nanoatrazine against susceptible weeds. *Front. Environ. Sci.* 6: 12.

89 Heckmann, L.H., Hovgaard, M.B., Sutherland, D.S. et al. (2011). Limit-test toxicity screening of selected inorganic nanoparticles to the earthworm *Eisenia fetida*. *Ecotoxicology* 20: 226–233.

90 Shoults-Wilson, W.A., Reinsch, B.C., Tsyusko, O.V. et al. (2011). Effect of silver nanoparticle surface coating on bioaccumulation and reproductive toxicity in earthworms (*Eisenia fetida*). *Nanotoxicology* 5: 432–444.

91 Unrine, J.M., Tsyusko, O.V., Hunyadi, S.E. et al. (2010). Effects of particle size on chemical speciation and bioavailability of copper to earthworms (*Eisenia fetida*) exposed to copper nanoparticles. *J. Environ. Qual.* 39: 1942–1953.

92 Oluah, M., Obiezue, R.N.N., Ochulor, A.J., and Onuoha, E. (2010). Toxicity and histopathological effect of atrazine (herbicide) on the earthworm Nsukkadrilus mbae under laboratory conditions. *Anim. Res. Int.* 7: 1287–1293.

93 Andrade, Ltd., do Espirito Santo Pereira, A., Fernandes Fraceto, L, and Bueno dos Reis Martinez, C. (2019). Can atrazine loaded nanocapsules reduce the toxic effects of this herbicide on the fish *Prochilodus lineatus*? A multi-biomarker approach. *Sci. Total Environ.* 663: 548–559.

94 Diez-Ortiz, M., Lahive, E., George, S. et al. (2015). Short-term soil bioassays may not reveal the full toxicity potential for nanomaterials; bioavailability and toxicity of silver ions ($AgNO_3$) and silver nanoparticles to earthworm Eisenia fetida in long-term aged soils. *Environ. Pollut.* 203: 191–198.

95 Tiwari, E., Mondal, M., Singh, N. et al. (2020). Effect of the irrigation water type and other environmental parameters on CeO_2 nanopesticide–clay colloid interactions. *Environ. Sci.: Processes Impacts* 22: 84–94.

96 Mahdavi, S., Jalali, M., and Afkhami, A. (2012). Removal of heavy metals from aqueous solutions using Fe_3O_4, ZnO, and CuO nanoparticles. *J. Nanopart. Res.* 14: 846–864.

97 Afkhami, A., Saber-Tehrani, M., and Bagheri, H. (2010). Simultaneous removal of heavy metal ions in wastewater samples using nano-alumina modified with 2,4-dinitrophenylhydrazine. *J. Hazard. Mater.* 181: 836–844.

98 Ahmadzadeh Tofighy, M. and Mohammadi, T. (2011). Adsorption of divalent heavy metals ions from water using carbon nanotube sheets. *J. Hazard. Mater.* 185: 140–147.

99 Rahmani, A., Zavvar Mosavi, H., and Fazli, M. (2010). Effect of nanostructured alumina on adsorption of heavy metals. *Desalination* 253: 94–100.

100 Navrotsky, A. (2000). Nanomaterials in the environment, agriculture, and technology (NEAT). *J. Nanopart. Res.* 2: 321–323.

101 Sharma, V.K. and Sohn, M. (2009). Aquatic arsenic: toxicity, speciation, transformations, and remediation. *Environ. Int.* 35: 743–759.

102 Simeonidis, K., Gkinis, T., Tresintsi, S. et al. (2011). Magnetic separation of hematite-coated Fe_3O_4 particles used as arsenic adsorbents. *Chem. Eng. J.* 168: 1008–1015.

103 Luo, X., Wang, C., Luo, S. et al. (2012). Adsorption of As (III) and As (V) from water using magnetite Fe_3O_4-reduced graphite oxide–MnO_2 nanocomposites. *Chem. Eng. J.* 187: 45–52.

104 Charlet, L., Morrin, G., Rose, J. et al. (2011). Reactivity at (nano) particle-water interfaces, redox processes, and arsenic transport in the environment. *Compt. Rend. Geosci.* 343: 123–139.

105 Ngomsik, A.F., Bee, A., Draye, M. et al. (2005). Magnetic nano and microparticles for metal removal and environmental applications: a review. *C.R. Chim.* 8: 963–970.

106 Chowdhury, S.R. and Yanful, E.K. (2010). Arsenic and chromium removal by mixed magnetite-maghemite nanoparticles and the effect of phosphate on removal. *J. Environ. Manage.* 91: 2238–2247.

107 Gupta, V.K. and Nayak, A. (2012). Cadmium removal and recovery from aqueous solutions by novel adsorbents prepared from orange peel and Fe_2O_3 nanoparticles. *Chem. Eng. J.* 180: 81–90.

108 Hu, J., Chen, G., and Lo, I.M.C. (2005). Removal and recovery of Cr(VI) from wastewater by maghemite nanoparticles. *Water Res.* 39: 4528–4536.

109 Liu, J.F., Zhao, Z.S., and Jiang, G.B. (2008). Coating Fe_3O_4 magnetic nanoparticles with humic acid for high efficient removal of heavy metals in water. *Environ. Sci. Technol.* 42: 6949–6954.

110 Huang, S.H. and Chen, D.H. (2009). Rapid removal of heavy metal cations and anions from aqueous solutions by an amino-functionalized magnetic nano adsorbent. *J. Hazard. Mater.* 163: 174–179.

111 Rick, A.R. and Arai, Y. (2010). Role of natural nanoparticles in phosphorus transport processes in Ultisols. *Soil Sci. Soc. Am. J.* 75: 335–347.

112 Heathwaite, L., Haygarth, P., Matthews, R. et al. (2005). Evaluating colloidal phosphorus delivery to surface waters from diffuse agricultural sources. *J. Environ. Qual.* 34: 287–298.

113 Kersting, A.B., Efurd, D.W., Finnegant, D.L. et al. (1999). Migration of plutonium in groundwater at the Nevada Test Site. *Nature* 397: 56–59.

114 Sen, T.K. and Khilar, K.C. (2006). Review on subsurface colloids and colloid associated contaminant transport in saturated porous media. *Adv. Colloid Interface Sci.* 119: 71–96.

115 Sharma, A., Hirata, T., Kawamoto, K. et al. (2010). Mobilization and transport of natural and water-dispersible colloids in repacked Okinawa red-yellow soil columns. *19th World Congress of Soil Science-Soil Solutions for a Changing World*, Brisbane, Australia (1–6 August 2010).

116 Johnson, W.P. and Amy, G.L. (1995). Facilitated transport and enhanced desorption of polycyclic aromatic hydrocarbons by natural organic matter in aquifer sediments. *Environ. Sci. Technol.* 29 (3): 807–817.

117 Mendive, C.B., Bredow, T., Feldhoff, A. et al. (2009). Adsorption of oxalate on anatase (100) and rutile (110) surfaces in aqueous systems: experimental results vs. theoretical predictions. *Phys. Chem. Chem. Phys.* 11 (11): 1794–1808.

118 Sparks, D.L. (2003). *Environmental Soil Chemistry*, 2e. San Diego, CA: Academic Press.

119 Huang, P.M. and Germida, J.J. (2002). Chemical and biological processes in the rhizosphere: metal pollutants. In: *Interactions Between Soil Particles and Microorganisms. Impact on the Terrestrial Ecosystem*, IUPAC Series on Analytical and Physical Chemistry of Environmental Systems, vol. 8 (eds. P.M. Huang, J.-M. Bollag and N. Senesi), 1–10. Chichester: Wiley.

120 Nowack, B. and Bucheli, T.D. (2007). Occurrence, behaviour and effects of nanoparticles in the environment. *Environ. Pollut.* 150: 5–12.

121 Koelmans, A.A., Jonker, M.T.O., Cornelissen, G. et al. (2006). Black carbon: the reverse of its dark side. *Chemosphere* 63: 365–377.

122 Qafoku, N.P. (2010). Terrestrial nanoparticles and their controls on soil–/geoprocesses and reactions. In: *Advances in Agronomy*, Chapter 2, vol. 107 (ed. D.L. Sparks), 33–91. San Diego, CA: Academic Press.

123 Crane, M. and Handy, R.D. (2007). An assessment of regulatory testing strategies and methods for characterizing the ecotoxicological hazards of nanomaterials. Report for Defra, London, UK. Retrieved from http://randd.defra.gov.uk/Document.aspx?DocumentID=2270.

124 Owen, R. and Handy, R.D. (2007). Formulating the problems for environmental risk assessment of nanomaterials. *Environ. Sci. Technol.* 41: 5582–5588.

125 Handy, R.D., Owen, R., and Valsami-Jones, E. (2008). The ecotoxicology of nanoparticles and nanomaterials: current status, knowledge gaps, challenges, and future needs. *Ecotoxicology* 17: 315–325.

126 Revell, P.A. (2006). The biological effects of nanoparticles. *Nanotechnol. Percept.* 2: 283–298.

127 Buzea, C., Pacheco, I., and Robbie, K. (2007). Nanomaterials and nanoparticles: sources and toxicity. *Biointerphases* 2: 17–71.

128 Kooter, I.M., Pennings, J.L.A., Opperhuizen, A., and Cassee, F.R. (2005). Gene expression pattern in spontaneously hypertensive rats exposed to urban particulate matter (EHC-93). *Inhal. Toxicol.* 17: 53–65.

129 Nel, A., Xia, T., Madler, L., and Li, N. (2006). Toxic potential of materials at ¨ the nanolevel. *Science* 311: 622–627.

130 Buffle, J. (2006). The key role of environmental colloids/nanoparticles for the sustainability of life. *Environ. Chem.* 3: 155–158.

131 Yang, L. and Watts, D.J. (2005). Particle surface characteristics may play an important role in phytotoxicity of alumina nanoparticles. *Toxicol. Lett.* 158: 122–132.

132 Moore, M.N. (2006). Do nanoparticles present ecotoxicological risks for the health‘ of the aquatic environment? *Environ. Int.* 32: 967–976.

133 Zhu, Y., Zhao, Q., Li, Y. et al. (2006). The interaction and toxicity of multi-walled carbon nanotubes with *Stylonychia mytilus*. *J. Nanosci. Nanotechnol.* 6: 1357–1364.

134 Gustafsson, O., Krusa, M., Zencak, Z. et al. (2009). Brown clouds over South Asia: biomass or fossil fuel combustion? *Science* 323 (5913): 495–498.

135 Engling, G. and Gelencser, A. (2010). Atmospheric brown clouds: from local air pollution to climate change. *Elements* 6 (4): 223–228.

136 Wilson, S.R., Solomon, K.R., and Tang, X. (2007). Changes in tropospheric composition and air quality due to stratospheric ozone depletion and climate change. *Photochem. Photobiol. Sci.* 6: 301.

137 Smita, S., Gupta, S.K., Bartonova, A. et al. (2012). Nanoparticles in the environment: assessment using the causal diagram approach. *Environ. Health* 11 (Suppl. 1): S13.

8

Nanotechnology: Advances in Plant and Microbial Science

CHAPTER MENU

8.1 Engineered Nanomaterials and Soil Remediation

Widespread scientific and industrial interest has been attracted by engineered nanomaterials (ENMs), illustrating prospects in the fields of material science and environmental science, medicine, agriculture, etc. [1–4]. The small-scale, high specific surface area, reactivity, and stability of ENMs allow them to remove recalcitrant pollutants and achieve target contaminant efficacy in complex environmental media [5]. By using ENMs, nanoscale structures, devices, and systems with novel properties and functionalities, nanotechnology provides innovative remediation solutions in pollutant reduction and resource reuse. Among the numerous environmental media, because of the variability and uncertainty of the soil–water interface, remediation of polluted soil and groundwater has shown significant technical and economic challenges. In situ control and successful mixing and productive interactions between the remediation agents and the target, pollutants are the challenges. Conventional ex situ remediation steps over the past two decades have

Nano-Technological Intervention in Agricultural Productivity, First Edition.
Javid A. Parray, Mohammad Yaseen Mir and Nowsheen Shameem.

included removing contaminated soils or recapping of fresh soils with identical contents or remedial techniques of engineering sort, including surfactant flushing, extraction of soil vapour, and thermal desorption. Emerging in situ clean-up technologies have recently been developed, such as chemical oxidation and reduction, sequestration, immobilization, and bioremediation. Innovation priorities in this field should include allowing, in addition to commonly targeted contaminants, simultaneous monitoring and consideration of emerging pollutants, such as heavy metals, organic compounds, polychlorinated biphenyls, polyfluoroalkyl substances, and other pollutants [6, 7].

The indigenous soil ecosystem is influenced by the anthropogenic introduction of ENMs to the environment [8]. There are growing questions about ENMs and soil remediation nanotechnology's health effects on the subsurface climate and soil ecosystem [9]. This influences the germination of plant seeds; the production of plant roots and shoots; and the growth of soil microorganisms and individual invertebrate animals such as snails, earthworms, or insect larvae that are affected in addition to pollution or likely entry into the food chain of groundwater or drinking water systems [4, 10, 11]. Therefore, the effects of ENMs on the soil environment, mostly those widely used for soil remediation during application processes, should not be overlooked.

A thorough evaluation of the applications and implications for soil remediation of ENMs and nanotechnology is reported as follows [12].

8.1.1 ENMs: Role in Soil Remediation

In situ technologies' underlying removal mechanisms include immobilization, photocatalytic degradation, oxidation similar to Fenton and Fenton, reduction reaction, and various other combinations [13, 14]. Typically, catalytic degradation and reduction reactions have also eliminated organic contaminants, while adsorption has played a central role in treating polluted heavy metals and metalloid soils. In soil remediation, ENMs that could simultaneously eliminate multiple contaminants by combined or sequential treatments tended to be of greater significance [15]. The presence of ENMs could promote both adsorption and redox degradation of micropollutants [15] in a simultaneous adsorption and oxidation scenario. In addition, the market for refractory contaminants and co-exposed pollutants to be mitigated by these multifunctional materials is an upward trend.

These ENMs include carbon nanomaterials (NMs), Fe_3O_4, TiO_2, ZnO, and nanoscale zero-valent iron (nZVI) nanoparticle nanocomposites. The ENMs are used to remediate soils contaminated by heavy metals and organic compounds.

8.1.1.1 Immobilization

The immobilization or adsorption method for removing pollutants, a reliable, cost-effective, and relatively environmentally friendly clean-up process for metal-contaminated soils, is currently used in the most significant number of remedial applications [16]. The selection of appropriate materials or coating compounds, depending on the various pollutants and soil conditions, is critical in immobilization remediation. To date, several nanosized additives are used in the immobilization of inorganic and organic contaminants in the soil matrix, such as carbon NMs, metal oxide NMs, nanocomposites, and other ENMs. Although carbon NMs absorb organic pollutants through van der Waals forces and π–π relationships, through surface complexation, metal oxide NMs and other nanocomposites immobilize heavy metals and organic compounds.

Carbon NMs are primarily based on their surface hydrophobicity, and broad adsorption capacity is used as adsorbents for the treatment of organic pollutants in water and soils [17]. The process and mechanism of adsorption and desorption of organic compounds on carbon NMs, in particular, CNTs, in water have been extensively studied to date [18, 19]. Because of the comparatively higher adsorption affinity of CNTs to these organic compounds than soil particles, the use of CNTs in soil remediation is gaining considerable traction [20]. Moreover, because of the low-barrier, charge-assisted hydrogen bonds and cation-π assisted π–π interactions, CNTs were documented to adsorb and sequester ionizable organic compounds [21] explicitly. The absorption by CNTs of polycyclic aromatic hydrocarbons (PAHs) (e.g. acenaphthylene, naphthalene, fluorene, phenanthrene, pyrene, etc.) has already been widely assessed under various environmental conditions [22]. The mineralization and mobilization of phenanthrene and benzo[a]pyrene could be inhibited by soil CNTs, thus reducing the extractability and bioavailability of these PAHs to plants and soil microorganisms [23]. CNTs can also act as an effective adsorbent for the immobilization of pyrene, especially when pretreated humic acid is added to ultrasound [20]. The affinity of CNTs to phenolic compounds for adsorption and ability depended on their physicochemical characteristics, including form, scale, number of layers, and number of functional groups of hydroxyls (–OH). The –OH replacement on the aromatic ring was able to boost the π–π interaction between the aromatic rings and CNTs, allowing CNTs to attract π–receptor and π–donor to the surface simultaneously, thus enhancing adsorption [23].

Metal Oxide NMs Fe_3O_4 and TiO_2 have great potential in the soil environment for the adsorption of different contaminants. Fe_3O_4 has a high capacity for heavy metals such as Cd and As [24, 25] to absorb and immobilize. Under conditions of 10–20 °C and pH 6.0, the highest Cd sorption efficiency of Fe_3O_4 occurred with a maximum adsorption potential of 37.03 mg/g [24, 26]. Surface modification using stabilizer compounds was generally appropriate because Fe_3O_4 particles appeared to agglomerate and settled out of the aqueous phase or delayed between soil particles in the absence of stabilizer [26]. Liang and Zhao [27] investigated the efficacy of starch-stabilized Fe_3O_4 for in situ sorption and immobilization of As in sandy loam soil [28]. Moreover, recent results have shown that coating with dissolved organic matter could significantly boost the sorption intensity of TiO_2 to PAHs, resulting in significant mitigation of the mobility of phenanthrene in the soil environment [29].

Nanocomposites and Other ENMs Nanocomposites have been documented to be more efficient in removing pollutants than single materials and have shown good efficiency after several desorption cycles of adsorption [30]. An and Zhao [31] developed a revolutionary class of Fe–Mn binary oxide NMs stabilized by polysaccharides and evaluated their capacity for sorption of As^{3+} in soils. The column elution results suggested that the application of Fe–Mn nanocomposites stabilized by carboxymethyl cellulose (CMC) almost completely converts soluble As^{3+} into the particle phase, significantly reducing its leachability by up to 96%. Mallampati et al. [32] synthesized a composite of nano-Fe/Ca/CaO/(PO_4) for dual separation and immobilization of Cs in polluted soils, achieving 100% immobilization of stable Cs (133Cs) under dry conditions only after simple grinding [33]. They also claimed that other ENMs, such as Zn(VI) and nanosized $MnFe_2O_4$ particles, exerted efficient adsorption of As, Hg, and Cr(VI) and showed great potential for in situ soil treatment [34, 35].

8.1.1.2 Photocatalytic Degradation

Photocatalytic degradation using UV-irradiated nano-photocatalysts or exposure to sunlight has been commonly used to degrade organic pollutants such as PAHs, polychlorinated biphenyls (PCBs), and pesticides in the treatment of water [36, 37]. It has recently been used for soil remediation [29, 38]. A traditional ex situ procedure is used to wash polluted soils with non-polar solvents and treat photocatalysts with leached water to remove toxic compounds [39]. Still, the in situ method involves the direct addition of ENMs to contaminated soils in conjunction with light irradiation [40]. Because of the soil's low light penetration, this approach is severely restricted and will involve continuous land ploughing [41]. Overall, the photocatalytic oxidation's efficiency was associated with soil properties, pH value, and the presence of organic matter. The number of hydroxyl radicals ($OH^{\cdot}$) formed by nano-photocatalysts in the soil environment was determined by these factors in the principle and played an essential role in the competitive sorption contaminants against soil particles [42]. TiO_2 is the most widely used ENM, degrading 78.3%, 23.4%, and 31.8% of pollutants under visible light penetration within five hours, respectively [42].

Fenton-like Reaction For the decontamination of organic wastewater, Fenton reagents ($H_2O_2 + Fe(II)$) have been widely used to oxidize organic compounds in the presence of ferrous iron (acting as catalysts) [43]. A growing number of studies have recently begun to probe its application in the subsurface environment [44]. As catalysts for underground Fenton-like oxidation procedures, many materials were used. The most common iron minerals and corresponding ENMs, such as Fe_3O_4 and iron-contained nanocomposites, have attracted considerable attention [45]. For the decontamination of organic wastewater, Fenton reagents ($H_2O_2 + Fe(II)$) have been widely used to oxidize organic compounds in the presence of ferrous iron (acting as catalysts) [43]. A growing number of studies have recently begun to probe its application in the subsurface environment [44]. As catalysts for underground Fenton-like oxidation procedures, many materials were used. The most common iron minerals and corresponding ENMs, such as Fe_3O_4 and iron-contained nanocomposites, have attracted considerable attention.

Reduction Reaction The mitigation of heavy metals (Cr(VI)) and organic compounds in polluted water, soil, and groundwater was successfully reported to minimize reactions [46]. In practical cases, nZVI, whether in the form of permeable reactive barriers or direct addition through injection wells, has already been applied on a pilot or field scale [47]. In most cases, nZVI could reduce Cr(VI) to Cr(III) because of strong reduction power and form complex precipitants such as ferrous chromite ($FeCr_2O_4$) [48]. The combination of nZVI with biochar or compost could increase the effectiveness of removing nZVI and Cr(VI) reaction operation by hindering accumulation and enhancing the disparity of iron particles or reducing the mixture's mobility soil [48]. For example, green tea modified nZVI (GT-nZVI) injection could reduce the soil-bound Cr(VI) in the soil column by up to 66% from the initial 100 mg/kg soil concentration [49, 50]. CMC-stabilized nZVI has also been reported to convert most exchangeable Cr to carbonate-bound and oxide-bound Fe–Mn, thereby reducing Cr's bioavailability and leachability in soil conditions [51]. The primary factors that could affect the degradation efficiency of nZVI to dichlorodiphenyltrichloroethane (DDT) are the properties of nZVI (e.g. reactivity, particle size, and suspension stability) and the soil pH condition [52].

Mixed Approaches and Multipurpose ENMs Multiple and mixed contaminants can be targeted simultaneously by a significant improvement in the design and synthesis of multifunctional ENMs and show more excellent contaminant selectivity among complex matrix components. The synergistic function of ENMs in adsorption (or chelation) and degradation concentrate and expose contaminants to a relatively high concentration of reactants, enhancing removal and/or conversion efficacy. For instance, electron transfer between ENMs and target compounds and free radical generation can both be facilitated by bringing contaminants to the surface of ENMs, thus improving surface-driven degradation processes. In water and wastewater treatment, surface oxidation has been promoted as an integral part of advanced oxidation processes, such as Fenton reaction and photo- and electrocatalytic oxidation processes, in which OH· is generated in situ on the catalyst or the electron surfaces. For example, the use of CNT-supported TiO_2 or single-atom Pt/SiC [53] as catalysts has been studies for the degradation of recalcitrant compounds, such as PFASs, through surface-mediated reactions. Besides, novel nanocomposites have been proposed for the single-step removal of organic and inorganic compounds and enhancing the stability of sequestrating efficacy under complex conditions [54]. While some of these exemplary ENMs require photoactivation and are therefore unsuitable for in situ applications, and some of them are water-based studies to date, they are of great potential for soil remediation when combined with other techniques being applied aboveground. An emerging potential to enhance underground emissions' remediation is to combine nanotechnology with bioremediation methods to achieve refractory compound degradation to minimize overall engineering costs [11, 55]. For example, SiO_2 NMs are of great potential to boost fertilizer mobility and bioaccessibility, thereby enhancing phytoremediation efficacy by using energy crops to extract PAHs from soils [56]. For example, during the remediation of contaminated metal and metalloid soils, nZVI, TiO_2, nanosized hydroxyapatite (nHAP), and some nanocomposites increase plants' extraction rate [11]. Fang et al. [57] suggested that the in situ degradation of 2,4-D by Fe_3O_4 is conducive to indigenous microbes. ENMs and nanotechnology have shown promising potential for soil and groundwater remediation, in which immobilization, photocatalytic degradation, Fenton-like oxidation, reduction reaction, etc., are the underlying mechanisms. To increase their treatment performance, the production of advanced ENMs covers the manufacture of multifunctional materials and the combination of other types of technologies.

8.2 Fate and Interactions of Nanomaterials in Soil

Soil is the largest recipient of NPs, so knowing the fate and behaviour of NPs in the soil becomes essential, as the soil is the ultimate sink for nanoparticles. The nanomaterial is exposed to various factors or agents, mainly plant roots and microorganisms, when it enters the soil, and establishes multiple relationships with them, most commonly synergistic or antagonistic. As primary particles and aggregates, the soil is often considered an externally significant environmental matrix, richest in natural NPs. The anthropogenic introduction of NPs into the soil could have a considerable influence on the soil biota. The main issues for the sustainable use of NPs in crop production are soil microbial biomass and diversity and root growth [58].

8.2.1 Nanoparticles and Plants

The nanoparticles or nanomaterials when applied to plants through foliar spray on soil application are ultimately absorbed by the plant foliage and/or roots and cause a complexity of effects. As nanoparticles are small in size, with a reasonable degree of free movement, they can join, translocate, and penetrate physiological barriers to travel within the plant tissues. Long-term NP effects are not entirely understood in plant systems, however. NPS can cause morphological, physiological, and genetic modifications that contribute to quantitative or qualitative plant growth changes. It is well known that when present in the soil, nanoparticles either penetrate plant tissues via root tips, rhizodermis, and lateral root junctions or when applied to the foliage via cuticles, trichomes, stomata, etc. Depending on the NP species and its concentration, nanoparticles' effects on plants can be growth-suppressive or protrusive.

8.2.2 Suppressive Effects on Plants

The nanoparticles can easily penetrate the root and cause toxic effects, restricting the expression of toxic symptoms in a group of cells and suppressing plant growth. In seed germination, root/shoot development, biomass production, transpiration photosynthesis, biomolecule synthesis, etc., nanoparticles' toxic effects on plants can appear [59]. Nanoparticles can cause plant growth to decline, degrade its quality, and lower the nutritional value of the product [60, 61]. NPS harms the growth and yield of crops and accumulates in plant tissues, fruits, grains, etc. A number of crops such as cucumber [62], coriander [63], lettuce [64, 65], onion [66], rice [67], mung bean [68], radish [69], spinach [70], soy [71, 72], tobacco [73], wheat [74], and mustard [75] have been shown to inhibit NPS seed germination, decrease plant growth, and decrease the photosynthetic rate and pigments. In Table 8.1 the toxic effects of NP on plants have been summarized.

8.2.3 Promontory Plant Effects

Some chemical nanoparticles, mainly those useful for plant growth, such as nitrogen, phosphorus, zinc, etc., stimulate plant growth and increase the production and yield of biomass. Nitrogen is an essential nutrient for plant growth and production, and because of leaching and other factors, 50–70% of the applied N dose is lost to the soil. Because of the high surface area-to-volume ratio [76], it has been found that the application of nano-nitrogen is much more efficient than even polymer-coated traditional slow-release N fertilizers. Likewise, P-fertilizers are also lost in surface waters because of the eutrophication issue [77]. Water-soluble phosphate salts are commercially available P fertilizers, readily dissolved in soil solution and available for plant uptake [78].

Healthy sources of P, on the other hand, such as naturally occurring phosphate rocks and apatites, have also been tried as P fertilizers where the phosphate is trapped in a tangible form and less readily accessible to plant roots and much less readily transportable by run-off or soil erosion [78]. However, in providing nutrient P to plants, these solids are less effective [78]. Besides, the large size of the particles is hampered by the application of stable phosphates in agriculture. However, nanosized apatite particles may be as efficient as

Table 8.1 Phytotoxic effect of nanoparticles on plants.

Nanoparticles	Plant species	Impacts	References
Silver (5–25 mm) 40 mg/l	*Psilaxis radiatus*	Growth inhibition	[60]
Silver (0.6–2 mm) 100 mg/l	Ryegrass, barley	Decreases seed germination and shoot length	[59]
Gold (25–100 mg/l)	*A. thaliana, Medicago sativa*	Retarded root length	[126]
Cerium oxide 500 mg/kg	*Raphanus sativus*	Retarded germination	[69]
Cerium oxide (200 mg/l) Lanthanum oxide (1000 mg/l)	*A. thaliana*	Oxidative stress and expression of genes	[61]
Cupric oxide 200–500 mg/l	*Vigna radiata*	Root length affected, chlorophyll and biomass content decreased	[68]
Cupric oxide 0–500 mg/l	*Brassica juncea*	Shoot and root growth decreased	[71]
Cupric oxide 500–1000 mg/l	Carrot	Decreased biomass	[190]
Iron 50–100 mg/l	Sunflower	Reduced uptake of nutrients and translocation	[191]
Iron 1–50 mg/l	Lettuce	Root elongation decreases	[192]
Nd_2O_3 100 mg/l	Pumpkin	Oxidative stress etc.	[193]
Zinc oxide 500–1500 mg/l	*Brassica campestris*	Reduced seed germination and seedling growth	[194]
Zinc oxide 0.05–0.50 mg/l	Soybean	Leaf chlorosis and growth retardant	[72]

widely used soluble P fertilizers in supplying the nutrient P but should reduce eutrophication. The application of nanosized hydroxyapatite in a greenhouse test increased the height of soybean plants by 30% compared with standard P fertilizer treatment. The treatment also increased the growth, dry biomass, and seed yield of soybean plants higher than standard P soybean plants [79].

About 50% of the world's arable soils have Zn deficiency [80]. Zn fertilizers' efficacy in providing Zn to plants in Zn-deficient soils largely depends on Zn's soil solubility. The most widely used Zn fertilizer [81] is inorganic ZnO. The most promising solution appears to be the use of ZnO NPs in the place of traditional ZnO, where a minimum dose will fulfil the Zn plant requirement [82]. By increasing the rate and extent of Zn dissolution, ZnO NPs will increase the fertilizer's efficiency and Zn availability to plants [83].

8.2.4 Nanoparticles and Impacts on Soil Microbes

In general, nanoparticles are considered to be harmful to microorganisms [84]. The accumulation of NPs in the soil will affect the microbial community and diversity adversely. The bioavailability of other toxins and nutrients can be affected explicitly by nanoparticles or indirectly by interactions with natural organic compounds [85]. Nanoparticles and their ions can create free radicals that can destroy the membrane of microbial cells, DNA, and mitochondria, leading to microbes' death [60, 61]. To determine nanoparticles' effect on microbial development, soil respiration, and enzymatic activity are essential determinants [86]. Metallic nanoparticles may impair enzymatic activities and bacterial abundance in the soil; soil types may also play an indispensable role in NP toxicity [87]. A danger to healthy microbial ecosystems can also be the introduction of NPS into natural environments. In flooded paddy soil, the TiO_2 and CuO NPs reduced soil microbial biomass and enzymatic activities and influenced their population structures [88]. Similar effects of ZnO, TiO_2, CeO_2, and Fe_3O_4 NPs on microbial enzymes such as invertase, urease, catalase, phosphatase, and bacterial communities in two different types of soils have been documented by You et al. [87]. The high Fe_3O_4 NP concentration significantly reduced the number of bacteria in the soil [89]. Chai et al. [90] observed decrease in the plate counts of Azotobacter, P-solubilizing and K-solubilizing bacteria and enzymatic activities due to ZnO and CeO_2 NPs. Similarly titanium dioxide NPs reduced the enzymatic activity and abundance and diversity of rhizoplane bacteria [91] by titanium dioxide NPs. Native soil bacteria have been toxic to CuO NPs, causing cavities, holes, membrane degradation, and lysis in the microbial cells [92].

As the Ag NP treatments substantially suppressed the colonization of both the fungi and the EC50 of Ag NPs was much smaller than the ionic Ag, the detailed impacts of individual nanoparticles on soil microbes are known. The reported Ag NP treatment resulted in a significant reduction in the colonization of two perennial ryegrass fungi [93, 94]. Safavi et al. [95] stated that nanosilver had excellent potential for removing bacterial contaminants during the tissue culture of the tobacco plant. Ag NPs/polyvinylpyrrolidone (PVP) exhibited antibacterial activity *Staphylococcus aureus*, *Escherichia coli*, *Pseudomonas aeruginosa*, and *Bacillus subtilis* spores [96]. Guzman et al. [97] Ag NPs showed healthy bactericidal activity against *E. coli*, *P. aeruginosa*, and *S. aureus*. The growth of *Alternaria alternata*, *Sclerotinia sclerotiorum*, *Macrophomina phaseolina*, and *Rhizoctonia solani* was significantly inhibited by silver nanoparticles at 15 mg/l [98]. The high surface area and the fraction of silver nanoparticles surface atoms have a high antimicrobial activity relative to the bulk of silver [99].

8.2.5 Zinc and Sulfur Nanoparticles

The nanoparticles of zinc oxide were also found to be more antimicrobial than the traditional ZnO powder. Compared with conventional ZnO, Tayel et al. [100] tested the efficacy of ZnO NPs against nine bacterial strains. The ZnO NPs are found to be more bacteria suppressive than powder. Treatment with ZnO NPs with *Salmonella typhimurium* and *S. aureus* resulted in 100% cell death within eight and four hours of application [100], respectively. Jayaseelan et al. [101] have confirmed the antibacterial activity of ZnO NPs against

P. aeruginosa. The antifungal efficacy of sulfur NPs (S NPs) was investigated in vitro against *Fusarium solani* and *Venturia inaequalis* [102]. The S-NPs (35 nm) were more effective in suppressing fungal colonization than the larger particles. The microscopic analysis showed that the leading cause of fungal suppression was the accumulation of S-NPs on the surface of the spores and hyphae and subsequent damage [102].

8.2.6 Copper and Silica Nanoparticles

Cioffi et al. [103] stated that *Botrytis cinerea* colonies more significant than copper salt powder were suppressed by Cu NPs. The Cu NP antifungal activities of polymer based copper nanocomposites against pathogenic fungi [103] and silica-silver nanoparticles against *B. cinerea*, *R. solani* and *Colletotrichum gloeosporioides* have been reported [104]. Copper nanoparticles showed efficient antimicrobial activity against Gram-positive and Gram-negative bacteria and fungi in soda-lime glass powder [105]. There have also been records of CuO nanoparticles' antibacterial activities against *S. aureus*, *B. subtilis*, *P. aeruginosa*, and *E. coli* [106].

8.3 Nanomaterials and Metal Components: Accumulation and Translocation Within Plants

The rapid growth of nanotechnology and the extensive commercial applications of ENMs, particularly the soil–plant system, increase the risk of discharge of engineered nanoparticles (ENPs) into the atmosphere. There are, in short, two critical routes by which ENPs can reach the soil–plant system. One route is to add sewage sludge to agriculture, mostly containing SiO_2, TiO_2, ZnO, and Ag NPs and the other is through the use of nano-agrochemicals such as nano-pesticides, nano-fertilizers, and nano-amendments, resulting in the direct incorporation into the soil–plant environment of SiO_2, TiO_2, Zn/ZnO, Fe/FeOx, Cu/CuO/$Cu(OH)_2$, CeO_2, and Ag NPs [107, 108].

After entering the soil–plant system, NPS would eventually interfere with plants, thereby potentially affecting plant physiology and probably food security. Nano-phytotoxicity is the most commonly studied component of studies related to plant-to-ENP interactions and the physiological, biochemical, and genetic impacts of both harmful and beneficial effects on plants [109]. In determining ENP phytotoxicity, the absorption, translocation, and accumulation of ENPs in plants play a critical role and may further affect human food safety [110]. The current knowledge of the absorption and translocation of ENPs in plants remains very limited and is not systematic. The numerous physiological barriers to the uptake and transport of ENPs in plants have been comprehensively studied by Schwab et al. [111]. The effect of plant physiology on plant uptake and transportation of ENPs, however, remains a concern. Another essential aspect is that ENPs are highly unstable and that abiotic or biotic transformations of ENPs can occur in the rhizosphere or within plants, such as redox reactions, aggregation, and dissolution ENPs, and this will significantly alter the bioavailability, toxicity, and fate of ENPs [112]. Different advanced analytical methods have been used in the past decade to identify the speciation and positioning of NPs at the tissue, cellular, and subcellular levels in organisms. However, there are still significant challenges in this area, particularly in quantitative analysis and in situ detection of NPs in complex matrices such as soils, plants, and other organisms [113].

8.3.1 NPS: Uptake and Translocation in Plants

Two different modes of plant exposure to NPs are available, namely, foliar exposure and root exposure. The cuticle is considered the first natural protection against NPs entering into the substrate tissue because the waxy cuticle protects higher plants' leaves against water loss and unregulated exchange of other solutes [114]. Via diffusion and permeation (lipophilic pathway) and polar solutes through polar aqueous pores (hydrophilic pathway) [115], there are two routes for the absorption of solutes across the cuticle. The size of the NPS below 4.8 nm will penetrate directly through the cuticular way through the cuticle. However, several studies have reported foliar absorption and accumulation of NPs larger than 5 nm. It is also worth noting that there can be life-stage changes in leaf cuticles, climatic conditions, and damage to pests and diseases, in some instances, the first barrier of cuticles against NPS [115].

Studies have demonstrated the uptake of hydrophilic substances through stomatal openings (stomatal path) in addition to the cuticular pathway. Stomatal apertures have a morphological scale of approximately 25 μm in length and 3–10 μm in width [115]. However, the actual size exclusion limit of stomatal aperture for NP penetration is still far from understood because of the distinctive geometric construction and physiological feature of stomata. Using an indirect calculation method, Eichert and Goldbach estimated the equivalent pore radius of this pathway to be greater than 20 nm [116]. The only confirmed route of foliar absorption of NPs from the leaf surface to the internal tissues is this stomatal pathway. Numerous studies support this NP uptake pathway, including the observation with transmission electron microscopy (TEM), confocal laser scanning microscopy (CLSM), or micro X-ray fluorescence (μ-XRF), of different NPs or their aggregates in leaf stomata and deeper tissues of various plant species including *Citrullus lanatus*, *Cucurbita pepo*, *Allium porrum*, *Lactuca sativa*, and *Arabidopsis thaliana* [117]. The SEL and lateral heterogeneity of the stomatal foliar uptake pathway for water-suspended fluorescent polystyrene NPs [115] were investigated by Eichert et al. CLSM uptake of 43 nm NPs through the stomatal pathway. In contrast, no uptake of 1.1 μm particles was observed. Once they entered the subtotal cavern, they also participated in distributing fluorescent NPs in the leaf apoplast. The experimental results indicate that Eichert and Goldbach's estimated stomatal pathway's pore radius was an underestimate. Kim et al. discovered that plasma membrane H^+ – ATPase activity was induced by nZVI and promoted the stomatal aperture opening. Besides, it is expected that plant species with different leaf morphologies, stomatal sizes, and densities will have various capacities for NP foliar uptake [118].

After entering the leaf apoplastic via the stomatal pathway, NPs can undergo long-distance transport through the vascular system. Photosynthate, sugars, and macromolecules in the leaf, including small RNA and proteins, can traditionally be transported downwards to the shoots and roots via the phloem system [119]. In general, the long-distance transport of liquids in higher plants occurs through the vascular system consisting of the conductive tissues of xylem and phloem. The direction of flow is from bottom to top in the xylem system (from root to shoot), while the flow direction is from top to bottom in the phloem system (from shoot to source). This plants' vascular system is non-circulatory, which implies that materials passing downward through the phloem's xylem do not flow back to their original sites. For NPS to translocate from leaf to root, the phloem system is the only possible foliar uptake pathway. Although foliar uptake of NPS has been established in many studies, no research has given direct evidence to support the subsequent translocation pathway of NPs to phloems in plants. Wang et al. studied the foliar uptake by watermelon of four metal oxide NPs of size 24–47 nm. They noticed that the small NPs were able to penetrate watermelon leaves using TEM along the stomatal pathway. NPS moved through the nodes and eventually reached the roots through the phloem sieve tubes [120]. Foliar absorption of CeO_2 NPs of primary size 8 ± 1 nm by cucumber was investigated by Hong et al. [121]. Inductively coupled plasma-optical emission spectrometry (ICP-OES) was identified by TEM in all treated CeO_2-NP plant tissues.

Such findings support the theory that NPS will travel up through the xylem and down through the phloem, although the NPS's transport between the xylem and the phloem remains uncertain. Subsequently, the NPS and its transformed products that are translocated to the roots may be exuded into the rhizosphere. The ones translocated in leaves can be percolated into the phyllosphere, which may further affect the microbial population in the rhizosphere or phyllosphere. Many symbiotic or pathogenic microorganisms in the phyllosphere are found in the natural environment [122]. Microorganisms of the phyllosphere can secrete extracellular polymeric substances (EPSs) or select chemicals as molecules of the quorum-sensing signal [123]. These exudates may cause NP dissolution and further absorption by plant leaves of metal ions, or they may form a mucous layer to protect leaf surfaces from NP absorption. Besides, at various life stages, the organ production levels for plants are somewhat different, which may also affect the foliar absorption of NPs. Undeveloped cuticles in newly growing leaves and flowers, for example, or natural senescence of old leaf cuticles can provide NPs with a high chance of entering the leaves. The leaves defensive systems, such as the cuticle, the epidermis, and even the mesophyll, can be damaged by some common leaf diseases such as leaf mould, blight, chlorosis, and necrosis [124].

8.3.2 NPS: Root Uptake and Translocation

Several factors such as particle size, surface functionalization, morphology, exposure conditions, plant organisms, plant growth stage, root integrity, and processes of the rhizosphere influence the plant absorption of NPs.

Particle size is one of the essential properties influencing plant root uptake of NPs. Sabo-Attwood et al. observed root absorption of 3.5 nm Au NPs by tobacco using μ-XRF and TEM, but 18 nm Au NPs remained agglomerated on the root's outer surfaces [125]. However, Taylor et al. found that *A. thaliana* roots did not specifically take up Au NPs (7–108 nm) [126]. Besides, Slomberg and Schoenfisch noted that SiO_2 NPs of up to 200 nm could be absorbed into the roots of *A. thaliana*. After all, fewer particles were considered to have increased particle size (14, 50, and 200 nm) [127]. Larue performed an elegant experiment to reveal the size-dependent uptake and transport of TiO_2 NPs in wheat using TEM and μ-XRF, which shows that NPS with primary diameters of less than 36 nm were accumulated in roots and distributed across the entire plant tissue without TEM and μ-XRF.

In comparison, NPS accumulated in wheat root parenchyma with primary diameters in the range of 36–140 nm but did not reach the stele and did not translate to the shoots. NPS greater than 140 nm did not accumulate in wheat roots [128]. However, in most experiments, the particle sizes used were based on the initial NPS's average height. Plants' uptake of NPS depends on the minimum, but not the average, size of NPs.

Another factor influencing the root uptake and translocation of NPs in plants is the surface charge. Generally, a border cell mucilage layer composed of negatively charged root secretions cover the plant root cap. Using X-ray computed nanotomography (nano-CT) and hyperspectral imaging microscopy, Avellan et al. found that mucilages secreted by *A. thaliana* roots were adsorbed to positively charged Au NPs (approximately 12 nm) and avoided the translocation of Au NPs into the root tissues. However, the mucilages were not adsorbed by negatively charged Au NPs (about 12 nm) and could translocate into the roots' apoplast [129]. Koelmel et al. also found that surface functionalization significantly influenced the root uptake and translocation of Au NPs (core diameter 2 nm) in rice using laser ablation–inductively coupled mass spectrometry (LA-ICPMS) (*Oryza sativa* L.). Accumulated Au concentrations in roots were as follows: Au NPs (+) > Au NPs (0) > Au NPs(−), while the reverse order was obtained in shoots, suggesting preferential translocation through the vascular system of negatively charged NPs [130, 131]. Because of the formation of nano-coronas by coatings of negatively charged root exudates on the surfaces of positively charged NPs, certain circumstances may reverse the surface charges of NPs. Like negatively charged NPs, these nano-coronas can be transported through the vascular system. More complex scenarios involve root exudates that may cause aggregation, partial dissolution, or conversion of NPs, further affecting their root absorption and bioavailability. This mechanism dominates the chemical properties of the NPS and the exudates as well as the microenvironment of the rhizosphere in the rhizosphere.

Researchers have focused on the effect on the absorption, accumulation, and transformation of NPs in the soil–plant system of physiological factors such as plant species, disease, and rhizospheric microorganisms [132]. Different plant organisms, possibly because of plant physiology and metabolic function variations, show various uptake capacities for NPs. For instance, in dicotyledons, a higher Ce translocation rate was observed than in monocotyledons [133–135]. It is useful for agricultural risk management to consider the species-dependent absorption of NP by plants. Mycorrhizal fungi can form symbiotic partnerships with higher plants and increase the plant root's functional surfaces [111]. This encourages water and nutrient uptake by plants and can affect the uptake and bioavailability of NPs. Studies show that when exposed to 2 nm Ag NPs, tomato plants colonized by mycorrhizal fungi accumulate lower Ag concentrations in their tissues than non-mycorrhizal plants. Still, when the plants were exposed to 15 nm Ag NPs [136], this impact was not evident.

8.3.3 Assimilated Root Uptake and Translocation Pathways of Nanoparticles

The role and composition of the cuticle of the root surface are similar to those of the cuticle of the leaf surface. Whether NPS can impregnate the root surface cuticle is still unknown. However, the cuticles on root hairs are underdeveloped, and the root tips of primary and secondary roots are developed. Therefore, NPS in these regions may be directly exposed to the epidermis [137]. In higher plants, there are two main pathways for root uptake and transport of NPs when approaching the root epidermis. The apoplastic pathway in which NPS first penetrates the cell wall's pores and then diffuses into the space between the cell wall and the plasma membrane, or passes through the intercellular area without crossing the cell membrane, has been proposed by most studies. Many studies have observed NPs or their aggregates using TEM or CLSM in the apoplastic root space, indicating the presence of an apoplastic pathway in plant roots, for example, 20 nm ZnO NPs, in ryegrass roots [129], 12 nm Au NPs in the roots of *A. thaliana* [129], 20–80 nm Ag NPs in the roots of *A. thaliana* [138], 43 nm CuO NPs in the roots of *Elsholtzia splendens* [139], and 22 nm La_2O_3 NPs in the roots of cucumbers [140]. Another possible explanation is that through diseased roots or physical wounding due to belowground herbivores and mechanical injuries, such as incidents during seedling transplantation, NPS reaches the intercellular spaces or even xylem [141]. NPS can move via the epidermis, cortex, and into the endodermis through the apoplastic pathway but are prevented by the Casparian strip, a belt of specialized cell wall material sealed by lipophilic hydrocarbons located around the vascular system in adjacent endodermis cells to avoid macromolecules and NPs from accessing the vascular cylinder [142]. However, NPS can reach the vascular system under certain special conditions and escape the Casparian strip through the root apoplast, such as the root tip region where the Casparian strip has not yet formed or the lateral root junction where the Casparian strip is disconnected [143, 144]. Experimental evidence of this pathway was given by Lv et al. who found that 30 nm ZnO NPs were collected at the lateral root junction of maize and within the attached xylem cylinder using TEM, μ-XRF, and fluorescence tracking [112]. Therefore, an essential apoplastic pathway through which NPs reach the vascular system may be the lateral root junction. Future studies are required to confirm whether this pathway occurs in other species of plants and specify the size of this pathway's size.

The symplastic road, which is a cell-to-cell transport pathway, is the other hypothesized pathway. There are two obstacles through which NPs can be transferred across the symplastic pathway, in addition to the crossing across the cell wall. One is cell membrane penetration through NPS and entry into the cytoplasm. The other is the transport of NPs by plasmodesmata (PD) to the neighbouring cells after entering the cell. Numerous theories for the transmembrane transport of NPs by plant cells have been suggested, e.g. via aquaporins, interconnected ion channels, endocytosis, and the broken membrane's intubation. Ion channel function for particular ions, however, and while aquaporins are non-selective, aquaporin protein channels are usually below 1 nm in diameter [144]. A water or ion channel pathway hypothesis for NP cellular internalization must face challenges from plant physiology's fundamental concepts [145]. The most feasible transmembrane pathway for NPs is by endocytosis, according to the literature. Compared with that of animal systems, plant cell endocytosis is poorly known. Recent experiments on NP internalization using isolated plant cells indicate that endocytosis pathways are involved. Onelli et al. for example, observed both clathrin-independent and clathrin-dependent pathways for tobacco protoplastic endocytosis of Au NPs using CLSM [146]. Torney et al. further reported that the surface properties of NPs play a crucial role in plant cell endocytosis, and there was endocytic absorption by tobacco mesophyll protoplasts of functionalized mesoporous silica nanoparticles (MSNs) of triethylene glycol (TEG) but still did not take up the functionalized MSN [147, 148].

Individual plant cells are divided by cell walls. Symplastic transport by plasmodesmata takes place intercellularly (PD). Plasmodesmata are channels that span the wall of the plant cell and allow intercellular communication between adjacent cells by linking the cytoplasm. Botanists have provided ample evidence that PD facilitates the cell-to-cell trafficking of biological macromolecules such as proteins, nucleic acids, RNA, and PD. The function of PD has been well reviewed as intercellular channels for macromolecular transport in plants [149]. In particular, non-targeted diffusion protein traffic in the range of approximately 3–20 kDa with a maximum size of roughly 3 nm [150] has been demonstrated to provide a model for intercellular NP traffic. In the first 1–2 mm of *Arabidopsis thaliana* root tips (plant root exposure to 20, 40, and 80 nm Ag NPs) [138], Geisler-Lee, using TEM, observed Ag-containing NPs. It aggregated clumps within root cell PD and middle lamella for the first time. TEM Au NPs or their aggregates in the root cell cytoplasm, cell walls, plastids, mitochondria, and particular PD of poplar plants (*Populus deltoides*) exposed to 15, 25, and 50 nm Au NPs have been observed by Zhai et al. [151]. To reveal the SEL of NPs for PD channels, microinjection strategies that have been used to evaluate protein transport through PD are worth pursuing.

8.3.4 NPS: Transformation in the Rhizosphere

Compared with their bulk counterparts, NPS are highly dynamic and unstable in the atmosphere because of their small-scale, high surface-to-volume ratio, and reactivity [152]. When NPS is released into the environment, several biotic and abiotic processes can occur, resulting in changes in their state of aggregation, surface chemistry, or speciation. Thus, the properties of pristine NPs cannot be preserved by NPs exposed to plants in the real world. In the past few years, various studies have centred on transforming NPs in soils, sediments, and wastewater and activated sludge [153]. For example, Ag NPs were found to convert to Ag_2S in sludge, converted 100 ZnO NPs to Zn_3 $I(PO_4)_2$, Fe oxy/hydroxides (Zn-FeOOH)-associated ZnS and Zn in sludge and biosolids [154, 155]. The rhizosphere is an area enriched with root exudates and microorganisms that are chemically and biologically active and can cause changes in NPS before their approach to root surfaces [156]. For example, Huang et al. found that Cu NPs and synthetic root exudate strong binding ability influenced Cu NPs' oxidation and reduction transformations to CuI(I) and Cu(II). They also significantly decreased Cu uptake and bioaccumulation in cucumber (*Cucumis sativus*) [157]. The dissolution of CuO NPs (~40 nm) in wheat (*Triticum aestivum*) rhizosphere soil was investigated by Gao et al. [158] and found that Cu increased in the readily available fraction ($CaCl_2$ extracted) and the labile fraction of Cu (diethylenetriaminepentaacetic acid [DTPA] extracted) decreased in the soil of the rhizosphere relative to the bulk soil [158]. The rhizosphere mechanism is caused by changes in soil pH, organic matter, mineral constituents, and the microbial community and causes them to change.

Furthermore, changes in these soil factors can lead to either opposite or adverse effects on NP-transformations [159]. Besides, most of the reactions are thermodynamically feasible but kinetically slow between NPs and soil matrices. Kinetics is thus one of the most important factors deciding the transformation of rhizospheric NPs [158]. The ageing process is also influenced by surface coatings and the particle size of NPs. Coutris et al. for example, have found that 20 nm uncoated Ag NPs are more resistant to ageing in the soil than Ag ions or 5 nm citrate-coated Ag NPs. During an ageing of 70 days, they can still serve as continuous sources of bioaccessible Ag ions [160]. Although related studies are still in their infancy, they remind us that aged soil NPS can vary considerably from their ionic shape, and this needs to be taken into account in the risk assessment of soil NPs. To explore the dissolution and transformation of NPs in soils, especially those representing high exposure risks to plants such as Cu-based NPs and ZnO, TiO_2, CeO_2, and Ag NPs, further field experiments, long-term studies, and systematic projects are urgently needed.

NP transformation at the rhizosphere's soil–root interface can affect the phytotoxicity and bioavailability of NPs [161]. Wang et al. studied the absorption, aggregation, and toxicity in cucumber and wheat of Ag_2S NPs. They found that 120 nm Ag_2S NPs were taken up by plant roots and subsequently delivered into leaf tissues as Ag_2S NPs without transformation [162, 163]. The effects of speciation and solubility of Cu-based NPs on Cu uptake and translocation in the *T. aestivum* roots were investigated by Spielman-Sun et al. Their findings revealed that $CuI(OH)_2$ high solubility NPs were rapidly absorbed by roots and further reduced and sulfided. Conversely, low-solubility CuO or CuS NPs were more persistent over the 48 hour post-exposure period, with as much as 80% of the non-transformed NPS [164].

8.4 Biotransformation of ENPs in Plants

In addition to transformation in the rhizosphere environments, NPS modifications have also been found to occur within or on plant tissues' surface, resulting in various accumulated elemental speciations in plants. Yin et al. discovered that Ag was partly oxidized in the root tissues of *Lolium multiflorum* for the first time. To explain this finding, two hypotheses were proposed: (i) direct root uptake of Ag NPs, followed by the oxidative transformation in root tissues, and (ii) dissolution of Ag NPs outside the root surface, followed by root uptake of ionic species [165]. Wang et al. established in *Chlamydomonas reinhardtii* the cellular internalization and intracellular biotransformation of NPs to Ag_2S and Ag-thiolates [166] but observed no similar transformation of Ag NPs within higher plant tissues. Au NPs are less active than Ag NPs, and the oxidative modification of Au NPs within plants has not been documented in any research.

In comparison, plants could minimize both Ag^+ and Au^+ to form NPs in plant roots and shoots [167]. It is a significant challenge to distinguish the pristine Ag or Au NPs from photosynthesized NPS, and measuring the values of Ag or Au isotope ratios can give some chance of determining them more clearly [167]. The most stable of the widely studied NPSs are metal oxide NPs such as TiO_2 and SiO_2 NPs and are present in plants in their new speciation [168]. On the other hand, the transformation of NPS such as ZnO, CuO, NiO, CeO_2, Yb_2O_3, and La_2O_3 has been reported, resulting in changes in accumulated plant speciation. The transformations of ZnO NPs exposed to various plants have been determined using synchrotron X-ray absorption spectroscopy (XAS) was used to determine the transformations of ZnO NPs exposed to different plants. Lv et al. found that most Zn accumulated in maize roots and shoots was under hydroponic exposure of ZnO NPs in forms such as Zn phosphate, primarily because of increased dissolution ZnO NPs in the rhizosphere and ionic [112] plant absorption and translocation of Zn. Related accumulated Zn speciation has also been observed in soil-grown wheat. Other forms of accumulated speciation have also been identified, such as soybean Zn-citrate and Zn-nitrate [169], soybean Zn-citrate, cowpea in-histidine and Zn-phytate [113], and velvet mesquite Zn-nitrate [170]. While distinct Zn speciation has been found in plants, a clear finding is that no ZnO was observed in shoots when exposed roots to ZnO NPs. Zn uptake, transport, and accumulation are predominantly in the form of Zn^{2+} released from ZnO NPs in plants. Exposure to CuO NPs and Cu^{2+} resulted in similar accumulated Cu speciation in wheat, with the exception that Cu(II) was reduced to Cu(I) within the plants for the CuO portion and Cu(I)-sulfur complexes were formed [171]. In soil-grown rice and maize, a similar reduction of Cu(II) to CuI(I) in plants was also observed [172]. Furthermore, Peng et al. found that 40 nm CuO NPs were transported to shoots from rice roots and dissolved Cu(II) was primarily combined with ligands of cysteine, citrate, and reduced phosphate and some of the Cu(II) to Cu_2O [173].

Rare earth oxide (REO) NPs, such as CeO_2 NPs, has been considered to be highly stable and do not transform the atmosphere and plants [174]. However, Zhang et al. have performed systematic studies on plant absorption and accumulation of REO NPs. In plants and the rhizosphere, they find the REO NPs such as CeO_2, Yb_2O_3, and La_2O_3 NPs. Zhang et al. reported hydroponic exposure of cucumber plants to 7 nm of CeO_2 NPs for 21 days, part of CeO_2 decreased to Ce(III) and formed needle-like $CePO_4$ clusters in intercellular regions and cucumber root epidermis, and during translocation to shoots formed Ce(III) complexes with carboxylates [175]. They hypothesized that organic acids facilitated the dissolution of CeO_2 NPs in root exudates, and the reduction of root exudates caused a reduction in the rhizosphere of Ce(IV) to Ce(III) [175]. In the intercellular regions of the cucumber roots, 22 nm La_2O_3 NPs or 15 nm Yb_2O_3 NPs exposed to hydroponically grown cucumbers were observed to be partially dissolved because of organic acids exuded from the roots and converted to needle-like $LaPO_4$ or $YbPO_4$ nanoclusters close to the dissolution and transformation processes of ZnO NPs [176].

8.5 Effect of Nanomaterials on Plants

8.5.1 Positive Effects

Plant growth conditions are related to the external environment and the absorption of nutrients. Some studies have shown that a significant impact on plants is the addition of nanomaterials to the soil. The addition of nano-TiO_2 can, for example, promote soybean growth by altering soybean chlorophyll or carotene, which is expected to increase the rate of soybean photosynthesis [177]. Similarly, adding 5 g/kg nHAP to soil can increase soil pH, fix Pb, increase the biomass of *Lolium perenne* L., and promote the growth of *L. perenne* L. [178]. The effects of nano-Fe_3O_4 on crops show that nano-Fe_3O_4 has no impact on *T. aestivum* germination rate, plant growth, and chlorophyll quality. After five days of hydroponic cultivation of *T. aestivum* L., however, plants with nano-Fe_3O_4 found that *T. aestivum* L. was improved by nano-Fe_3O_4 [179].

Moreover, studies have shown that the addition of a particular concentration of nano-CeO_2 to soil increases the biomass of *T. aestivum* L. in greenhouse conditions. It also raises the height of the plant [180]. Besides, nanomaterials have a specific effect on soil remediation, too. In general, nanomaterials cover a broad specific surface area; they have microporous network-structured channels and have a reasonably high degree of adsorption capacity [181]. These characteristics provide them with a particular effect of adsorption, leaving unregulated plant growth in soil polluted with heavy metals. The exposure time to nanomaterials is an influencing factor in the development of plants [182]. Phytoremediation may be used for a limited period [183].

Consequently, with the systematic use of plants and nanomaterials, emission remediation consists primarily of two factors. One consists of influencing plants' inner structure with various nanomaterials such as chlorophyll and carotenoid to encourage plant growth. Another consists of using nanomaterials to absorb heavy metals or metal-like metals in the soil together with plants. Because nanomaterials have positive effects on crops and promote crop growth, it will be useful to use them at reasonable concentrations in agricultural production.

8.5.2 Toxicity

Detrimental effects of nanomaterials on biology [184] and, in particular, nanomaterial toxicity in plants have been documented, and carbon nanomaterials have been found to inhibit the absorption of nutrients from plants [185]. The nanomaterials hinder plant protein content, chlorophyll, carotenoids and biomass. Nanomaterials can also extend the harvest time of crop *T. aestivum* L. Nanoparticles are passed from plant roots to plant leaves and join the food chain [186], amid plants being able to excrete potentially toxic elements.

Nano-CeO_2 alters the amino acids' content and increases the grain protein levels and *T. aestivum* L. growth. Different concentrations of nanomaterials after treatment with nanomaterials had different effects on other *Glycine max* sections (L.) Merry., and total chlorophyll content was inversely proportional to nano-ZnO's attention [72]. nZVI has a specific inhibitory effect on rice carotenoids and rice chlorophyll, affecting rice photosynthesis, thereby decreasing plant biomass [187]. There is a distinct morphology of nZVI in the soil, so the effect is different. On rice seedlings, it has an inhibitory effect [187]. Nano-CeO_2 may also have a specific effect on crop growth, in addition to crop quality. Nano-CeO_2 will promote plant growth at low concentrations and boost levels of biomass and chlorophyll content. It can also extend the *T. aestivum* L. harvesting time. The addition of 500 mg/kg nano-ZnO to soil contributes to accumulating a significant amount of Zn in *G. max* (L.) Mar. plants, thus preventing *G. max* (L.) Merr. [188]. Although the toxicity of individual elements can be prevented by plants, the metal components of nanomaterial are still transferred from plant roots to air-contact plant parts and edible leaves, thus entering the food chain [186]. There are the same disadvantages of synthetic materials. Artificially generated nano-ZnO can be transferred to overground plant components, which have a specific impact on the biomass of plant leaves, resulting in a decrease in the dry weight of plant leaves [72]. Researchers have evaluated that nanomaterials impact on crops through the measurement of leaf transpiration, stomatal conductance, and net photosynthesis. Corresponding findings show that nanomaterials significantly reduce the fresh weight of plants and their average biomass, and when the concentration was 50 mg/l, root elongation was considerably reduced [72]. High-oxidized nanomaterial concentrations have unique toxic effects, resulting in plant growth retardation, growth inhibition, etc. The absorption by the leaves and roots of water and nutrient plants indicates that different parts of plants are subject to toxicity levels [189].

References

1 Javed, I., Peng, G., Xing, Y. et al. (2019). Inhibition of amyloid beta toxicity in zebrafish with a chaperone-gold nanoparticle dual strategy. *Nat. Commun.* 10 (1): 3780.

2 Xu, M., Zhao, T., Wang, C. et al. (2019). High power density tower-like triboelectric nanogenerator for harvesting arbitrary directional water wave energy. *ACS Nano* 13 (2): 1932–1939.

3 Wang, P. (2018). Emerging investigator series: the rise of nano-enabled photothermal materials for water evaporation and clean water production by sunlight. *Environ. Sci. Nano* https://doi.org/10.1039/c8en00156a.

4 Chang, J., Zhang, L., and Wang, P. (2018). Intelligent environmental nanomaterials. *Environ. Sci. Nano* 5: 811–836.

5 Zhang, T., Lowry, G.V., Capiro, N.L. et al. (2019). In situ remediation of subsurface contamination: opportunities and challenges for nanotechnology and advanced materials. *Environ. Sci. Nano* 6 (5): 1283–1302.

6 Iavicoli, I., Leso, V., Beezhold, D.H., and Shvedova, A.A. (2017). Nanotechnology in agriculture: opportunities, toxicological implications, and occupational risks. *Toxicol. Appl. Pharmacol.* 329: 96–111.

7 Gong, X., Huang, D., Liu, Y. et al. (2018). Remediation of contaminated soils by biotechnology with nanomaterials: bio-behavior, applications, and perspectives. *Crit. Rev. Biotechnol.* 38 (3): 1–14.

8 Shrivastava, M., Srivastav, A., Gandhi, S. et al. (2019). Monitoring of engineered nanoparticles in soil-plant system: a review. *Environ. Nanotechnol. Monit. Manag.*

9 O'Connor, D. and Hou, D. (2018). Targeting cleanups towards a more sustainable future. *Environ. Sci. Process. Impacts.*

10 Lin, D. and Xing, B. (2007). Phytotoxicity of nanoparticles: inhibition of seed germination and root growth. *Environ. Pollut.*

11 Zhu, Y., Xu, F., Liu, Q. et al. (2019). Nanomaterials and plants: positive effects, toxicity and the remediation of metal and metalloid pollution in soil. *Sci. Total Environ.* 662: 414–421.

12 Lin, S., Yu, T., Yu, Z. et al. (2018). Nanomaterials safer-by-design: an environmental safety perspective. *Adv. Mater.* https://doi.org/10.1002/adma.201705691.

13 Deng, R., Lin, D., Zhu, L. et al. (2017). Nanoparticle interactions with co-existing contaminants: joint toxicity, bioaccumulation and risk. *Nanotoxicology* 5: 591–612.

14 Cecchin, I., Reddy, K.R., Thomé, A. et al. (2017). Nanobioremediation: integration of nanoparticles and bioremediation for sustainable remediation of chlorinated organic contaminants in soils. *Int. Biodeterior. Biodegrad.* 119: 419–428.

15 Trujillo-Reyes, J., Peralta-Videa, J.R., and Gardea-Torresdey, J.L. (2014). Supported and unsupported nanomaterials for water and soil remediation: are they a useful solution for worldwide pollution? *J. Hazard Mater.* 280: 487–503.

16 Barzegar, G., Jorfi, S., Soltani, R.D.C. et al. (2017). Enhanced Sono-Fenton-like oxidation of PAH-contaminated soil using nano-sized magnetite as catalyst: optimization with response surface methodology. *Soil Sediment Contam.* 26 (5): 538–557.

17 Gupta, V.K. and Saleh, T.A. (2013). Sorption of pollutants by porous carbon, carbon nanotubes and fullerene – an overview. *Environ. Sci. Pollut. Res.* 20 (5): 2828–2843.

18 Yang, K. and Xing, B. (2007). Desorption of polycyclic aromatic hydrocarbons from carbon nanomaterials in water. *Environ. Pollut.* 145 (2): 529–537.

19 Peng, H., Zhang, D., Pan, B., and Peng, J. (2017). Contribution of hydrophobic effect to the sorption of phenanthrene, 9-phenanthrol and 9,10-phenanthrenequinone on carbon nanotubes. *Chemosphere* 168: 739–747.

20 Zhang, W., Lu, Y., Sun, H. et al. (2019). Effects of multi–walled carbon nanotubes on pyrene adsorption and desorption in soils: the role of soil constituents. *Chemosphere.*

21 Kah, M., Sigmund, G., Xiao, F., and Hofmann, T. (2017). Sorption of ionizable and ionic organic compounds to biochar, activated carbon and other carbonaceous materials. *Water Res.*

22 Li, S., Turaga, U., Shrestha, B. et al. (2013). Mobility of polyaromatic hydrocarbons (PAHs) in soil in the presence of carbon nanotubes. *Ecotoxicol. Environ. Saf.*

23 Wu, W., Jiang, W., Zhang, W. et al. (2013). Influence of functional groups on desorption of organic compounds from carbon nanotubes into water: insight into desorption hysteresis. *Environ. Sci. Technol.*

24 Sebastian, A., Nangia, A., and Prasad, M.N.V. (2019). Cadmium and sodium adsorption properties of magnetite nanoparticles synthesized from *Hevea brasiliensis* Muell. Arg. bark: relevance in amelioration of metal stress in rice. *J. Hazard. Mater.*

25 Sun, J., Chillrud, S.N., Mailloux, B.J. et al. (2016). Enhanced and stabilized arsenic retention in microcosms through the microbial oxidation of ferrous iron by nitrate. *Chemosphere.*

26 Liang, Q., An, B., and Zhao, D. (2013). Removal and immobilization of arsenic in water and soil using polysaccharide-modified magnetite nanoparticles. In: *Monitoring Water Quality: Pollution Assessment, Analysis, and Remediation*, 1e (ed. S. Ahuja), 285–297. Elsevier.

27 Liang, Q. and Zhao, D. (2014). Immobilization of arsenate in a sandy loam soil using starch-stabilized magnetite nanoparticles. *J. Hazard. Mater.*

28 Tan, X., Wang, X., Chen, C., and Sun, A. (2007). Effect of soil humic and fulvic acids, pH and ionic strength on Th(IV) sorption to TiO_2 nanoparticles. *Appl. Radiat. Isot.*

29 Wang, X., Ma, E., Shen, X. et al. (2014). Effect of model dissolved organic matter coating on sorption of phenanthrene by TiO_2 nanoparticles. *Environ. Pollut.*

30 Mahmoud, M.E., Osman, M.M., Yakout, A.A., and Abdelfattah, A.M. (2018). Water and soil decontamination of toxic heavy metals using aminosilica-functionalized-ionic liquid nanocomposite. *J. Mol. Liq.*

31 An, B. and Zhao, D. (2012). Immobilization of As(III) in soil and groundwater using a new class of polysaccharide stabilized Fe–Mn oxide nanoparticles. *J. Hazard. Mater.*

32 Mallampati, S.R., Mitoma, Y., Okuda, T. et al. (2015). Solvent-free synthesis and application of nano-Fe/Ca/CaO/[PO_4] composite for dual separation and immobilization of stable and radioactive cesium in contaminated soils. *J. Hazard. Mater.* 297: 74–82.

33 Sun, R.J., Chen, J.H., Fan, T.T. et al. (2018). Effect of nanoparticle hydroxyapatite on the immobilization of Cu and Zn in polluted soil. *Environ. Sci. Pollut. Res.*

34 Eyvazi, B., Jamshidi-Zanjani, A., and Khodadadi Darban, A. (2019). Immobilization of hexavalent chromium in contaminated soil using nano-magnetic $MnFe_2O_4$. *J. Hazard. Mater.*

35 Azari, P. and Bostani, A.A. (2017). Reducing As availability in calcareous soils using nanoscale zero valent iron. *Environ. Sci. Pollut. Res.*

36 Zhu, Y., Liu, X., Hu, Y. et al. (2019). Behavior, remediation effect and toxicity of nanomaterials in water environments. *Environ. Res.*

37 Osin, O.A., Yu, T., Cai, X. et al. (2018). Photocatalytic degradation of 4-nitrophenol by C, N-TiO_2: degradation efficiency vs. embryonic toxicity of the resulting compounds. *Front. Chem.*

38 Rachna, R.M. and Shanker, U. (2019). Degradation of tricyclic polyaromatic hydrocarbons in water, soil and river sediment with a novel TiO_2 based heterogeneous nanocomposite. *J. Environ. Manage.*

39 Fenoll, J., Flores, P., Hellín, P. et al. (2014). Minimization of methabenzthiazuron residues in leaching water using amended soils and photocatalytic treatment with TiO_2 and ZnO. *J. Environ. Sci. (China).*

40 Zhang, L., Li, P., Gong, Z., and Li, X. (2008). Photocatalytic degradation of polycyclic aromatic hydrocarbons on soil surfaces using TiO_2 under UV light. *J. Hazard. Mater.*

41 Gu, J., Dong, D., Kong, L. et al. (2012). Photocatalytic degradation of phenanthrene on soil surfaces in the presence of nanometer anatase TiO_2 under UV-light. *J. Environ. Sci. (China).*

42 Chang Chien, S.W., Chang, C.H., Chen, S.H. et al. (2011). Effect of sunlight irradiation on photocatalytic pyrene degradation in contaminated soils by micro-nano size TiO_2. *Sci. Total Environ.*

43 Hou, L., Wang, L., Royer, S., and Zhang, H. (2016). Ultrasound-assisted heterogeneous Fenton-like degradation of tetracycline over a magnetite catalyst. *J. Hazard. Mater.*

44 Usman, M., Hanna, K., and Haderlein, S. (2016). Fenton oxidation to remediate PAHs in contaminated soils: a critical review of major limitations and counter-strategies. *Sci. Total Environ.*

45 Garrido-Ramírez, E.G., Theng, B.K.G., and Mora, M.L. (2010). Clays and oxide minerals as catalysts and nanocatalysts in Fenton-like reactions – a review. *Appl. Clay Sci.*

46 Zhao, X., Liu, W., Cai, Z. et al. (2016). An overview of preparation and applications of stabilized zero-valent iron nanoparticles for soil and groundwater remediation. *Water Res.*

47 Song, Y., Fang, G., Zhu, C. et al. (2019). Zero-valent iron activated persulfate remediation of polycyclic aromatic hydrocarbon-contaminated soils: an in situ pilot-scale study. *Chem. Eng. J.*

48 Qian, L., Liu, S., Zhang, W. et al. (2019). Enhanced reduction and adsorption of hexavalent chromium by palladium and silicon rich biochar supported nanoscale zero-valent iron. *J. Colloid Interface Sci.*

49 Chrysochoou, M., Johnston, C.P., and Dahal, G. (2012). A comparative evaluation of hexavalent chromium treatment in contaminated soil by calcium polysulfide and green-tea nanoscale zero-valent iron. *J. Hazard. Mater.*

50 Franco, D.V., Da Silva, L.M., and Jardim, W.F. (2009). Reduction of hexavalent chromium in soil and ground water using zero-valent iron under batch and semi-batch conditions. *Water Air Soil Pollut.*

51 Zhang, R., Zhang, N., and Fang, Z. (2018). In situ remediation of hexavalent chromium contaminated soil by CMC-stabilized nanoscale zero-valent iron composited with biochar. *Water Sci. Technol.*

52 El-Temsah, Y.S., Oughton, D.H., and Joner, E.J. (2013). Effects of nano-sized zero-valent iron on DDT degradation and residual toxicity in soil: a column experiment. *Plant Soil.*

53 Huang, D., De Vera, G.A., Chu, C. et al. (2018). Single-atom Pt catalyst for effective C—F bond activation via hydrodefluorination. *ACS Catal.*

54 Liu, J., Cheng, S., Cao, N. et al. (2019). Actinia-like multifunctional nanocoagulant for single-step removal of water contaminants. *Nat. Nanotechnol.*

55 Romeh, A.A. and Ibrahim Saber, R.A. (2020). Green nano-phytoremediation and solubility improving agents for the remediation of chlorfenapyr contaminated soil and water. *J. Environ. Manage.*

56 Włóka, D., Placek, A., Smol, M. et al. (2019). The efficiency and economic aspects of phytoremediation technology using *Phalaris arundinacea* L. and *Brassica napus* L. combined with compost and nano SiO_2 fertilization for the removal of PAH's from soil. *J. Environ. Manage.*

57 Fang, J., Shan, X.Q., Wen, B. and Huang, R.X. (2013). Mobility of TX100 suspended multiwalled carbon nanotubes (MWCNTs) and the facilitated transport of phenanthrene in real soil columns. *Geoderma* 207–208: 1–7.

58 Khan, M.R., Ahamad, F., and Rizvi, T.F. (2019). Effect of nanoparticles on plant pathogens. In: *Advances in Phytonanotechnology* (eds. M. Ghorbanpour and S.H. Wani), 215–240. London: Academic Press.

59 El-Temsah, Y.S. and Joner, E.J. (2012). Impact of Fe and Ag nanoparticles on seed germination and differences in bioavailability during exposure in aqueous suspension and soil. *Environ. Toxicol.* 27: 42–49.

60 Lee, W.M., Kwak, J.I., and An, Y.J. (2012). Effect of silver nanoparticles in crop plants *Phaseolus radiates* and *Sorghum bicolor*: media effect on phytotoxicity. *Chemosphere* 86: 491–499.

61 Ma, C., Liu, H., Guo, H. et al. (2016). Defense mechanisms and nutrient displacement in *Arabidopsis thaliana* upon exposure to CeO_2 and In_2O_3 nanoparticles. *Environ. Sci. Nano* 3: 1369–1379.

62 Moon, Y.S., Park, E.S., Kim, T.O. et al. (2014). SELDI-TOF MS-based discovery of a biomarker in *Cucumis sativus* seeds exposed to CuO nanoparticles. *Environ. Toxicol. Pharmacol.* 38: 922–931.

63 Zuverza-Mena, N., Medina-Velo, I.A., Barrios, A.C. et al. (2015). Copper nanoparticles/compounds impact agronomic and physiological parameters in cilantro (*Coriandrum sativum*). *Environ. Sci. Process Impacts* 17: 1783–1793.

64 Trujillo-Reyes, J., Majumdar, S., Botez, C.E. et al. (2014). Exposure studies of core–shell Fe/Fe_3O_4 and Cu/CuO NPs to lettuce (*Lactuca sativa*) plants: are they a potential physiological and nutritional hazard. *J. Hazard Mater.* 267: 255–263.

65 Hong, J., Rico, C.M., Zhao, L. et al. (2015). Toxic effects of copper-based nanoparticles or compounds to lettuce (*Lactuca sativa*) and alfalfa (*Medicago sativa*). *Environ. Sci. Process Impacts* 17: 177–185.

66 Cvjetko, P., Milošić, A., Domijan, A.M. et al. (2017). Toxicity of silver ions and differently coated silver nanoparticles in *Allium cepa* roots. *Ecotoxicol. Environ. Saf.* 137: 18–28.

67 Shaw, A.K. and Hossain, Z. (2013). Impact of nano-CuO stress on rice (*Oryza sativa* L.) seedlings. *Chemosphere* 93: 906–915.

68 Nair, P.M.G., Kim, S.H., and Chung, I.M. (2014). Copper oxide nanoparticle toxicity in mung bean (*Vigna radiata* L.) seedlings: physiological and molecular level responses of in vitro grown plants. *Acta Physiol. Plant* 36: 2947–2958.

69 Corral-Diaz, B., Peralta-Videa, J.R., Alvarez-Parrilla, E. et al. (2014). Cerium oxide nanoparticles alter the antioxidant capacity but do not impact tuber ionome in *Raphanus sativus* (L). *Plant Physiol. Biochem.* 84: 277–285.

70 Singh, D. and Kumar, A. (2016). Impact of irrigation using water containing CuO and ZnO nanoparticles on *Spinach oleracea* grown in soil media. *Bull. Environ. Contam. Toxicol.* 97: 548–553.

71 Nair, P.M.G. and Chung, I.M. (2015). A mechanistic study on the toxic effect of copper oxide nanoparticles in soybean (*Glycine max* L.) root development and lignification of root cells. *Biol. Trace Elem. Res.* 162: 342–352.

72 Priester, J.H., Moritz, S.C., Espinosa, K. et al. (2017). Damage assessment for soybean cultivated in soil with either CeO_2 or ZnO manufactured nanomaterials. *Sci. Total Environ.* 579: 1756–1768.

73 Frazier, T.P., Burklew, C.E., and Zhang, B. (2014). Titanium dioxide nanoparticles affect the growth and microRNA expression of tobacco (*Nicotiana tabacum*). *Funct. Integr. Genomics* 14: 75–83.

74 Vannini, C., Domingo, G., Onelli, E. et al. (2014). Phytotoxic and genotoxic effects of silver nanoparticles exposure on germinating wheat seedlings. *J. Plant Physiol.* 171: 1142–1148.

75 Nair, P.M.G. and Chung, I.M. (2015). Study on the correlation between copper oxide nanoparticles induced growth suppression and enhanced lignification in Indian mustard (*Brassica juncea* L.). *Ecotoxicol. Environ. Saf.* 113: 302–313.

76 De Rosa, M.C., Monreal, C., Schnitzer, M. et al. (2010). Nanotechnology in fertilizers. *Nat. Nanotechnol.* 5: 91.

77 Conley, D.J., Paerl, H.W., Howarth, R.W. et al. (2009). Controlling eutrophication: nitrogen and phosphorus. *Science* 323: 1014–1015.

78 Fageria, N.K. (2009). *The Use of Nutrients in Crop Plants*. Boca Raton, FL: CRC Press.

79 Liu, R. and Lal, R. (2014). Synthetic apatite nanoparticles as a phosphorus fertilizer for soybean (*Glycine max*). *Sci. Rep.* 4: 5686.

80 Sillanpaa, M. (1990). *Micronutrient Assessment at the Country Level: An International Study*, 208. Rome: FAO.

81 Martens, D.C. and Westermann, D.T. (1991). Fertilizer applications for correcting micronutrient deficiencies. In: *Micronutrients in agriculture* (eds. J.J. Mortvedt, F.R. Cox, L.M. Shuman and R.M. Welch), 549–592. Madison, WI: Soil Science Society of America.

82 Khan, M.R. and Rizvi, T.F. (2017). Application of nanofertilizer and nanopesticides for improvements in crop production and protection. In: *Nanoscience and Plant–Soil Systems* (eds. M. Ghorbanpour, M. Khanuja and A. Varma), 405–427. Cham: Springer.

83 Panwar, J. (2012). Positive effect of zinc oxide nanoparticles on tomato plants: a step towards developing nano-fertilizers 01/2012. *Proceeding of 3rd International Conference on Environmental Research and Technology (ICERT)*, Penang (May 30–June 1 2012).

84 Khan, M.R., Adam, V., Rizvi, T.F. et al. (2019). Nanoparticle–plant interactions: two-way traffic. *Small* 15: 1901794.

85 Haris, Z. and Ahmad, I. (2017). Impact of metal oxide nanoparticles on beneficial soil microorganisms and their secondary metabolites. *Int. J. Life-Sci. Sci. Res.* 3: 1020–1030.

86 Simonin, M. and Richaume, A. (2015). Impact of engineered nanoparticles on the activity, abundance, and diversity of soil microbial communities: a review. *Environ. Sci. Pollut. Res.* 22: 13710–13723.

87 You, T., Liu, D., Chen, J. et al. (2018). Effects of metal oxide nanoparticles on soil enzyme activities and bacterial communities in two different soil types. *J. Soils Sediments* 18: 211–221.

88 Xu, C., Peng, C., Sun, L. et al. (2015). Distinctive effects of TiO_2 and CuO nanoparticles on soil microbes and their community structures in flooded paddy soil. *Soil Biol. Biochem.* 86: 24–33.

89 Cao, J., Feng, Y., Lin, X., and Wang, J. (2016). Arbuscular mycorrhizal fungi alleviate the negative effects of iron oxide nanoparticles on bacterial community in rhizospheric soils. *Front. Environ. Sci.* 4 (10).

90 Chai, H., Yao, J., Sun, J. et al. (2015). The effect of metal oxide nanoparticles on functional bacteria and metabolic profiles in agricultural soil. *Bull. Environ. Contam. Toxicol.* 94: 490–495.

91 Buzea, C., Pacheco, I.I., and Robbie, K. (2007). Nanomaterials and nanoparticles: sources and toxicity. *Biointerphases* 2: MR17–MR71.

92 Concha-Guerrero, S.I., Brito, E.M.S., Piñón-Castillo, H.A. et al. (2014). Effect of CuO nanoparticles over isolated bacterial strains from agricultural soil. *J. Nanomater.* 2014: 206.

93 Jo, Y.K., Kim, B.H., and Jung, G. (2009). Antifungal activity of silver ions and nanoparticles on phytopathogenic fungi. *Plant Dis.* 93: 1037–1043.

94 Kasprowicz, M.J., Kozioł, M., and Gorczyca, A. (2010). The effect of silver nanoparticles on phytopathogenic spores of *Fusarium culmorum*. *Can. J. Microbiol.* 56: 247–253.

95 Safavi, K., Mortazaeinezahad, F., Esfahanizadeh, M., and Javad Asgari, M. (2011). In vitro antibacterial activity of nanomaterial for using in tobacco plants tissue culture. *World Acad. Sci. Eng. Technol.* 79: 372–373.

96 Bryaskova, R., Pencheva, D., Nikolov, S., and Kantardjiev, T. (2011). Synthesis and comparative study on the antimicrobial activity of hybrid materials based on silver nanoparticles (AgNps) stabilized by polyvinylpyrrolidone (PVP). *J. Chem. Biol.* 4: 185–191.

97 Guzman, M.G., Dille, J., and Godet, S. (2009). Synthesis of silver nanoparticles by chemical reduction method and their antibacterial activity. *Int. J. Chem. Biol. Eng.* 2: 104–111.

98 Krishnaraj, C., Ramachandran, R., Mohan, K., and Kalaichelvan, P.T. (2012). Optimization for rapid synthesis of silver nanoparticles and its effect on phytopathogenic fungi. *Spectrochim. Acta, Part A* 93: 95–99.

99 Singh, S., Singh, B.K., Yadav, S.M., and Gupta, A.K. (2015). Applications of nanotechnology in agricultural and their role in disease management. *Res. J. Nanosci. Nanotechnol.* 5: 1–5.

100 Tayel, A.A., El-Tras, W.F., Moussa, S. et al. (2011). Antibacterial action of zinc oxide nanoparticles against foodborne pathogens. *J. Food Saf.* 31: 211–218.

101 Jayaseelan, C., Rahuman, A.A., Kirthi, A.V. et al. (2012). Novel microbial route to synthesize ZnO nanoparticles using *Aeromonas hydrophila* and their activity against pathogenic bacteria and fungi. *Spectrochim. Acta, Part A* 90: 78–84.

102 Rao, K.J. and Paria, S. (2013). Use of sulfur nanoparticles as a green pesticide on *Fusarium solani* and *Venturia inaequalis* phytopathogens. *RSC Adv.* 3: 10471–10478.

103 Cioffi, N., Torsi, L., Ditaranto, N. et al. (2004). Antifungal activity of polymer-based copper nanocomposite coatings. *Appl. Phys. Lett.* 85: 2417–2419.

104 Park, H.J., Kim, S.H., Kim, H.J., and Choi, S.H. (2006). A new composition of nanosized silica-silver for control of various plant diseases. *Plant Pathol. J.* 22: 295–302.

105 Esteban-Tejeda, L., Malpartida, F., Esteban-Cubillo, A. et al. (2009). Antibacterial and antifungal activity of a soda-lime glass containing copper nanoparticles. *Nanotechnology* 20: 505701.

106 Azam, A., Ahmed, A.S., Oves, M. et al. (2012). Size-dependent antimicrobial properties of CuO nanoparticles against gram-positive and-negative bacterial strains. *Int. J. Nanomed.* 7: 3527–3535.

107 Chhipa, H. (2017). Nanofertilizers and nanopesticides for agriculture. *Environ. Chem. Lett.* 15: 15–22.

108 Ma, X.M., Geiser-Lee, J., Deng, Y., and Kolmakov, A. (2010). Interactions between engineered nanoparticles (ENPs) and plants: phytotoxicity, uptake and accumulation. *Sci. Total Environ.* 408: 3053–3061.

109 Zhao, L.J., Hu, Q.R., Huang, Y.X. et al. (2017). Activation of antioxidant and detoxification gene expression in cucumber plants exposed to a $Cu(OH)_2$ nanopesticide. *Environ. Sci. Nano* 4: 1750–1760.

110 Tripathi, D.K., Tripathi, A., Singh Shweta, S. et al. (2017). Uptake, accumulation and toxicity of silver nanoparticle in autotrophic plants, and heterotrophic microbes: a concentric review. *Front. Microbiol.* 8: 07.

111 Schwab, F., Zhai, G.S., Kern, M. et al. (2016). Barriers, pathways and processes for uptake, translocation and accumulation of nanomaterials in plants-critical review. *Nanotoxicology* 10: 257–278.

112 Lv, J.T., Zhang, S.Z., Luo, L. et al. (2015). Accumulation, speciation and uptake pathway of ZnO nanoparticles in maize. *Environ. Sci. Nano* 2: 68–77.

113 Wang, P., Menzies, N.W., Lombi, E. et al. (2013). Fate of ZnO nanoparticles in soils and cowpea (*Vigna unguiculata*). *Environ. Sci. Technol.* 47: 13822–13830.

114 Pollard, M., Beisson, F., Li, Y.H., and Ohlrogge, J.B. (2008). Building lipid barriers: biosynthesis of cutin and suberin. *Trends Plant Sci.* 13: 236–246.

115 Eichert, T., Kurtz, A., Steiner, U., and Goldbach, H.E. (2008). Size exclusion limits and lateral heterogeneity of the stomatal foliar uptake pathway for aqueous solutes and water suspended nanoparticles. *Physiol. Plant.* 134: 151–160.

116 Eichert, T. and Goldbach, H.E. (2008). Equivalent pore radii of hydrophilic foliar uptake routes in stomatous and astomatous leaf surfaces-further evidence for a stomatal pathway. *Physiol. Plant.* 132: 491–502.

117 Larue, C., Castillo-Michel, H., Sobanska, S. et al. (2014). Foliar exposure of the crop *Lactuca sativa* to silver nanoparticles: evidence for internalization and changes in Ag speciation. *J. Hazard. Mater.* 264: 98–106.

118 Uzu, G., Sobanska, S., Sarret, G. et al. (2010). Foliar lead uptake by lettuce exposed to atmospheric fallouts. *Environ. Sci. Technol.* 44: 1036–1042.

119 Lough, T.J. and Lucas, W.J. (2006). Integrative plant biology: role of phloem long-distance macromolecular trafficking. *Annu. Rev. Plant Biol.* 57: 203–232.

120 Wang, W.N., Tarafdar, J.C., and Biswas, P. (2013). Nanoparticle synthesis and delivery by an aerosol route for watermelon plant foliar uptake. *J. Nanopart. Res.* 15: 1417.

121 Hong, J., Peralta-Videa, J.R., Rico, C. et al. (2014). Evidence of translocation and physiological impacts of foliar applied CeO_2 nanoparticles on cucumber (*Cucumis sativus*) plants. *Environ. Sci. Technol.* 48: 4376–4385.

122 Vorholt, J.A. (2012). Microbial life in the phyllosphere. *Nat. Rev. Microbiol.* 10: 828–840.

123 Lindow, S.E. and Brandl, M.T. (2003). Microbiology of the phyllosphere. *Appl. Environ. Microbiol.* 69: 1875–1883.

124 Sankaran, S., Mishra, A., Ehsani, R., and Davis, C. (2010). A review of advanced techniques for detecting plant diseases. *Comput. Electron. Agric.* 72: 1–13.

125 Sabo-Attwood, T., Unrine, J.M., Stone, J.W. et al. (2012). Uptake, distribution and toxicity of gold nanoparticles in tobacco (*Nicotiana xanthi*) seedlings. *Nanotoxicology* 6: 353–360.

126 Taylor, A.F., Rylott, E.L., Anderson, C.W., and Bruce, N.C. (2014). Investigating the toxicity, uptake, nanoparticle formation and genetic response of plants to gold. *PLoS One* 9: e93793.

127 Slomberg, D.L. and Schoenfisch, M.H. (2012). Silica nanoparticle phytotoxicity to *Arabidopsis thaliana*. *Environ. Sci. Technol.* 46: 10247–10254.

128 Larue, C., Laurette, J., Herlin-Boime, N. et al. (2012). Accumulation, translocation and impact of TiO_2 nanoparticles in wheat (*Triticum aestivum* spp.): influence of diameter and crystal phase. *Sci. Total Environ.* 431: 197–208.

129 Avellan, A., Schwab, F., Masion, A. et al. (2017). Nanoparticle uptake in plants: gold nanomaterial localized in roots of *Arabidopsis thaliana* by X-ray computed nanotomography and hyperspectral imaging. *Environ. Sci. Technol.* 51: 8682–8691.

130 Koelmel, J., Leland, T., Wang, H.H. et al. (2013). Investigation of gold nanoparticles uptake and their tissue level distribution in rice plants by laser ablation-inductively coupled-mass spectrometry. *Environ. Pollut.* 174: 222–228.

131 Spielman-Sun, E., Lombi, E., Donner, E. et al. (2017). Impact of surface charge on cerium oxide nanoparticle uptake and translocation by wheat (*Triticum aestivum*). *Environ. Sci. Technol.* 51: 7361–7368.

132 Keller, A.A., Adeleye, A.S., Conway, J.R. et al. (2017). Comparative environmental fate and toxicity of copper nanomaterials. *Nanoimpact* 7: 28–40.

133 Layet, C., Auffan, M., Santaella, C. et al. (2017). Evidence that soil properties and organic coating drive the phytoavailability of cerium oxide nanoparticles. *Environ. Sci. Technol.* 51: 9756–9764.

134 Glenn, J.B., White, S.A., and Klaine, S.J. (2012). Interactions of gold nanoparticles with freshwater aquatic macrophytes are size and species dependent. *Environ. Toxicol. Chem.* 31: 194–201.

135 Judy, J.D., Unrine, J.M., Rao, W. et al. (2012). Bioavailability of gold nanomaterials to plants: importance of particle size and surface coating. *Environ. Sci. Technol.* 46: 8467–8474.

136 Noori, A., White, J.C., and Newman, L.A. (2017). Mycorrhizal fungi influence on silver uptake and membrane protein gene expression following silver nanoparticle exposure. *J. Nanopart. Res.* 19: 66.

137 Lin, D.H. and Xing, B.S. (2008). Root uptake and phytotoxicity of ZnO nanoparticles. *Environ. Sci. Technol.* 42: 5580–5585.

138 Geisler-Lee, J., Wang, Q., Yao, Y. et al. (2013). Phytotoxicity, accumulation and transport of silver nanoparticles by *Arabidopsis thaliana*. *Nanotoxicology* 7: 323–337.

139 Shi, J., Peng, C., Yang, Y. et al. (2014). Phytotoxicity and accumulation of copper oxide nanoparticles to the Cu-tolerant plant *Elsholtzia splendens*. *Nanotoxicology* 8: 179–188.

140 Ma, Y.H., He, X., Zhang, P. et al. (2011). Phytotoxicity and biotransformation of La_2O_3 nanoparticles in a terrestrial plant cucumber (*Cucumis sativus*). *Nanotoxicology* 5: 743–753.

141 Dietz, K.J. and Herth, S. (2011). Plant nanotoxicology. *Trends Plant Sci.* 16: 582–589.

142 Geldner, N., Roppolo, D., De Rybel, B. et al. (2011). A novel protein family mediates Casparian strip formation in the endodermis. *Nature* 473: 380–384.

143 Schymura, S., Fricke, T., Hildebrand, H., and Franke, K. (2017). Elucidating the role of dissolution in CeO_2 nanoparticle plant uptake by smart radiolabeling. *Angew. Chem. Int. Ed.* 56: 7411–7414.

144 McCully, M. (1995). How do real roots work-some new views of root structure. *Plant Physiol.* 109: 1–6.

145 Luu, D.T. and Maurel, C. (2005). Aquaporins in a challenging environment: molecular gears for adjusting plant water status. *Plant Cell Environ.* 28: 85–96.

146 Onelli, E., Prescianotto-Baschong, C., Caccianiga, M., and Moscatelli, A. (2008). Clathrin-dependent and independent endocytic pathways in tobacco protoplasts revealed by labelling with charged nanogold. *J. Exp. Bot.* 59: 3051–3068.

147 Etxeberria, E., Gonzalez, P., Baroja-Fernandez, E., and Romero, J.P. (2006). Fluid phase endocytic uptake of artificial nanospheres and fluorescent quantum dots by sycamore cultured cells. *Plant Signaling Behav.* 1: 196–200.

148 Li, H., Ye, X., Guo, X. et al. (2016). Effects of surface ligands on the uptake and transport of gold nanoparticles in rice and tomato. *J. Hazard. Mater.* 314: 188–196.

149 Zambryski, P. and Crawford, K. (2000). Plasmodesmata: gatekeepers for cell-to-cell transport of developmental signals in plants. *Annu. Rev. Cell Dev. Biol.* 16: 393–421.

150 Oparka, K.J. and Turgeon, R. (1999). Sieve elements and companion cells-traffic control centers of the phloem. *Plant Cell* 11: 739–750.

151 Zhai, G.S., Walters, K.S., Peate, D.W. et al. (2014). Transport of gold nanoparticles through plasmodesmata and precipitation of gold ions in woody poplar. *Environ. Sci. Technol. Lett.* 1: 146–151.

152 Auffan, M., Rose, J., Bottero, J.Y. et al. (2009). Towards a definition of inorganic nanoparticles from an environmental, health and safety perspective. *Nat. Nanotechnol.* 4: 634–641.

153 Lowry, G.V., Espinasse, B.P., Badireddy, A.R. et al. (2012). Long-term transformation and fate of manufactured Ag nanoparticles in a simulated large scale freshwater emergent wetland. *Environ. Sci. Technol.* 46: 7027–7036.

154 Lombi, E., Donner, E., Tavakkoli, E. et al. (2012). Fate of zinc oxide nanoparticles during anaerobic digestion of wastewater and post-treatment processing of sewage sludge. *Environ. Sci. Technol.* 46: 9089–9096.

155 Mitrano, D.M., Motellier, S., Clavaguera, S., and Nowack, B. (2015). Review of nanomaterial aging and transformations through the life cycle of nano-enhanced products. *Environ. Int.* 77: 132–147.

156 Seshadri, B., Bolan, N.S., and Naidu, R. (2015). Rhizosphere-induced heavy metal(loid) transformation in relation to bioavailability and remediation. *J. Soil Sci. Plant Nutr.* 15: 524–548.

157 Huang, Y., Zhao, L., and Keller, A.A. (2017). Interactions, transformations, and bioavailability of nano-copper exposed to root exudates. *Environ. Sci. Technol.* 51: 9774–9783.

158 Gao, X., Avellan, A., Laughton, S. et al. (2018). CuO nanoparticle dissolution and toxicity to wheat (*Triticum aestivum*) in rhizosphere soil. *Environ. Sci. Technol.* 52: 2888–2897. https://doi.org/10.1021/acs.est.7b05816.

159 Nowack, B. and Bucheli, T.D. (2007). Occurrence, behavior and effects of nanoparticles in the environment. *Environ. Pollut.* 150: 5–22.

160 Coutris, C., Joner, E.J., and Oughton, D.H. (2012). Aging and soil organic matter content affect the fate of silver nanoparticles in soil. *Sci. Total Environ.* 420: 327–333.

161 Rico, C.M., Johnson, M.G., and Marcus, M.A. (2018). Cerium oxide nanoparticles transformation at the root-soil interface of barley (*Hordeum vulgare* L.). *Environ. Sci. Nano* 5: 1807–1812.

162 Wang, P., Lombi, E., Sun, S.K. et al. (2017). Characterizing the uptake, accumulation and toxicity of silver sulfide nanoparticles in plants. *Environ. Sci. Nano* 4: 448–460.

163 Stegemeier, J.P., Schwab, F., Colman, B.P. et al. (2015). Speciation matters: bioavailability of silver and silver sulfide nanoparticles to alfalfa (*Medicago sativa*). *Environ. Sci. Technol.* 49: 8451–8460.

164 Spielman-Sun, E., Lombi, E., Donner, E. et al. (2018). Temporal evolution of copper distribution and speciation in roots of *Triticum aestivum* exposed to CuO, $Cu(OH)_2$, and CuS nanoparticles. *Environ. Sci. Technol.* 52: 9777–9784.

165 Yin, L.Y., Cheng, Y.W., Espinasse, B. et al. (2011). More than the ions: the effects of silver nanoparticles on *Lolium multiflorum*. *Environ. Sci. Technol.* 45: 2360–2367.

166 Wang, S., Lv, J., Ma, J., and Zhang, S. (2016). Cellular internalization and intracellular biotransformation of silver nanoparticles in *Chlamydomonas reinhardtii*. *Nanotoxicology* 10: 1129–1135.

167 Lu, D., Liu, Q., Zhang, T. et al. (2016). Stable silver isotope fractionation in the natural transformation process of silver nanoparticles. *Nat. Nanotechnol.* 11: 682–686.

168 Larue, C., Khodja, H., Herlin-Boime, N. et al. (2011). Investigation of titanium dioxide nanoparticles toxicity and uptake by plants. *J. Phys. Conf. Ser.* 304: 012057.

169 Lopez-Moreno, M.L., de la Rosa, G., Hernandez-Viezcas, J.A. et al. (2010). Evidence of the differential biotransformation and genotoxicity of ZnO and CeO_2 nanoparticles on soybean (*Glycine max*) plants. *Environ. Sci. Technol.* 44: 7315–7320.

170 Hernandez-Viezcas, J.A., Castillo-Michel, H., Servin, A.D. et al. (2011). Spectroscopic verification of zinc absorption and distribution in the desert plant *Prosopis juliflora-velutina* (velvet mesquite) treated with ZnO nanoparticles. *Chem. Eng. J.* 170: 346–352.

171 Dimkpa, C.O., McLean, J.E., Latta, D.E. et al. (2012). CuO and ZnO nanoparticles: phytotoxicity, metal speciation, and induction of oxidative stress in sand-grown wheat. *J. Nanopart. Res.* 14: 1125.

172 Peng, C., Xu, C., Liu, Q. et al. (2017). Fate and transformation of CuO nanoparticles in the soil-rice system during the life cycle of rice plants. *Environ. Sci. Technol.* 51: 4907–4917.

173 Peng, C., Duan, D., Xu, C. et al. (2015). Translocation and biotransformation of CuO nanoparticles in rice (*Oryza sativa* L.) plants. *Environ. Pollut.* 197: 99–107.

174 Birbaum, K., Brogioli, R., Schellenberg, M. et al. (2010). No evidence for cerium dioxide nanoparticle translocation in maize plants. *Environ. Sci. Technol.* 44: 8718–8723.

175 Zhang, P., Ma, Y.H., Zhang, Z.Y. et al. (2012). Biotransformation of ceria nanoparticles in cucumber plants. *ACS Nano* 6: 9943–9950.

176 Zhang, P., Ma, Y.H., Zhang, Z.Y. et al. (2012). Comparative toxicity of nanoparticulate/bulk Yb_2O_3 and $YbCl_3$ to cucumber (*Cucumis sativus*). *Environ. Sci. Technol.* 46: 1834–1841.

177 Singh, J. and Lee, B.-K. (2018). Effects of nano-TiO_2 particles on bioaccumulation of 133 Cs from the contaminated soil by soybean (*Glycine max*). *Process Saf. Environ. Prot.* 116: 301–311.

178 Ding, L., Li, J., Liu, W. et al. (2017). Influence of nano-hydroxyapatite on the metal bioavailability, plant metal accumulation and root exudates of ryegrass for phytoremediation in lead-polluted soil. *Int. J. Environ. Res. Public Health* 14: 532.

179 Iannone, M.F., Groppa, M.D., de Sousa, M.E. et al. (2016). Impact of magnetite iron oxide nanoparticles on wheat (*Triticum aestivum* L.) development: evaluation of oxidative damage. *Environ. Exp. Bot.* 131: 77–88.

180 Du, W., Gardea-Torresdey, J.L., Ji, R. et al. (2015). Physiological and biochemical changes imposed by CeO_2 nanoparticles on wheat: a life cycle field study. *Environ. Sci. Technol.* 49: 11884–11893.

181 Yang, Z., Fang, Z., Zheng, L. et al. (2016). Remediation of lead contaminated soil by biochar-supported nano-hydroxyapatite. *Ecotoxicol. Environ. Saf.* 132: 224–230.

182 Romeh, A.A.A. (2018). Green silver nanoparticles for enhancing the phytoremediation of soil and water contaminated by Fipronil and degradation products. *Water Air Soil Pollut.*: 229.

183 Huang, D.-L., Chen, G.-M., Zeng, G.-M. et al. (2015). Synthesis and application of modified zero-valent iron nanoparticles for removal of hexavalent chromium from wastewater. *Water Air Soil Pollut.*: 226.

184 Chen, M., Qin, X., and Zeng, G. (2017). Biodegradation of carbon nanotubes, graphene, and their derivatives. *Trends Biotechnol.* 35: 836–846.

185 Chen, M., Zhou, S., Zhu, Y. et al. (2018). Toxicity of carbon nanomaterials to plants, animals and microbes: recent progress from 2015–present. *Chemosphere* 206: 255–264.

186 Vittori Antisari, L., Carbone, S., Bosi, S. et al. (2018). Engineered nanoparticles effects in soil-plant system: basil (*Ocimum basilicum* L.) study case. *Appl. Soil Ecol.* 123: 551–560.

187 Wang, J., Fang, Z., Cheng, W. et al. (2016). Higher concentrations of nanoscale zero-valent iron (nZVI) in soil induced rice chlorosis due to inhibited active iron transportation. *Environ. Pollut.* 210: 338–345.

188 Rico, C.M., Barrios, A.C., Tan, W. et al. (2015). Physiological and biochemical response of soil-grown barley (*Hordeum vulgare* L.) to cerium oxide nanoparticles. *Environ. Sci. Pollut. Res. Int.* 22: 10551–10558.

189 Song, U. and Lee, S. (2016). Phytotoxicity and accumulation of zinc oxide nanoparticles on the aquatic plants *Hydrilla verticillata* and *Phragmites australis*: leaf-type-dependent responses. *Environ. Sci. Pollut. Res. Int.* 23: 8539–8545.

190 Ebbs, S.D., Bradfield, S.J., Kumar, P. et al. (2016). Accumulation of zinc, copper, or cerium in carrot (*Daucus carota*) exposed to metal oxide nanoparticles and metal ions. *Environ. Sci. Nano* 3: 114–126.

191 Martínez-Fernández, D. and Komárek, M. (2016). Comparative effects of nanoscale zero-valent iron (nZVI) and Fe_2O_3 nanoparticles on root hydraulic conductivity of *Solanum lycopersicum* L. *Environ. Exp. Bot.* 131: 128–136.

192 Liu, R., Zhang, H., and Lal, R. (2016). Effects of stabilized nanoparticles of copper, zinc, manganese, and iron oxides in low concentrations on lettuce (*Lactuca sativa*) seed germination: nanotoxicants or nanonutrients. *Water Air Soil Pollut.* 227: 42.

193 Chen, G., Ma, C., Mukherjee, A. et al. (2016). Tannic acid alleviates bulk and nanoparticle Nd_2O_3 toxicity in pumpkin: a physiological and molecular response. *Nanotoxicology* 10: 1243–1253.

194 Zafar, H., Ali, A., and Zia, M. (2017). CuO nanoparticles inhibited root growth from *Brassica nigra* seedlings but induced root from stem and leaf explants. *Appl. Biochem. Biotechnol.* 181: 365–378.

9

Food Application and Processing: Nanotechniques and Bioactive Delivery Systems

CHAPTER MENU

9.1 Introduction

As medications against various diseases, humans have commonly used plant-based natural products. Based on traditional knowledge and procedures, modern medicines are predominantly derived from herbs. Natural resources account for 25% of the substantial pharmaceutical compounds and their available derivatives [1, 2]. A base for the development of new drugs is a natural compound of various molecular backgrounds. A recent development in the discovery of natural product-based drugs has been interested in the design of synthetically suitable lead molecules that imitate their counterparts' structure [3]. Natural products display impressive characteristics, such as excellent chemical diversity, macromolecular-specific chemical and biological properties, and less toxicity. This renders them conducive to the development of novel drugs [4]. Computational studies have also helped to envisage molecular drug interactions and establish drug technologies for the next decade, such as target-based drug discovery and drug delivery.

Because of many benefits, pharmaceutical firms are reluctant to invest further in drug development and drug delivery systems based on natural products [5]. Instead, they exploit the libraries of chemical compounds available to discover novel drugs. However, several major diseases, including cancer, diabetes, cardiovascular, inflammatory, and microbial diseases, are now being tested for natural compounds. This is mainly because there are specific benefits to natural medicines, such as reduced toxicity and side effects, low costs,

Nano-Technological Intervention in Agricultural Productivity, First Edition.
Javid A. Parray, Mohammad Yaseen Mir and Nowsheen Shameem.

and potential therapeutics. Considerations about biocompatibility and toxicity of natural substances, however, present a more critical problem when it comes to using them as medicine. As a result, many natural compounds do not clear the clinical trial phases [6–8]. The use of large-scale drug delivery materials presents significant challenges, including in vivo volatility, poor bioavailability, poor solubility, poor body absorption, target-specific delivery problems, and the possible adverse drug effects of tonic efficacy. Using new drug delivery systems to direct drugs to particular parts of the body could be an alternative that could solve these crucial problems [9, 10]. Therefore, nanotechnology plays an important role, with immense success, in advanced medicine/drug formulations, targeting the arena, and controlled release and delivery of drugs. It has been shown that nanotechnology crosses the barriers between biological and physical sciences by applying nanostructures and nanophases in various fields of science [11], in particular, in nanomedicine and nano-based drug delivery systems where such particles are of considerable interest [12, 13]. Nanomaterials can be well characterized as a material with sizes ranging from 1 to 100 nm that affect nanomedicine's frontiers, starting from biosensors, microfluidics, drug delivery, and tissue engineering microarray tests [14–16]. Nanotechnology employs curative agents at the nanoscale stage for nanomedicines. Nanoparticles have been powered by the field of biomedicine involving nanobiotechnology, drug delivery, biosensors, and tissue engineering [17]. Because nanoparticles comprise materials engineered at the atomic or molecular level, they are typically nanospheres of small size [18]. Compared to more detailed products, they can also pass more quickly in the human body. There are unique structural, chemical, mechanical, magnetic, electrical, and biological properties of nanoscale-sized particles.

Nanomedicines have become well recognized in recent days because nanostructures could be used as delivery agents by encapsulating drugs or adding therapeutic drugs and delivering them more precisely with a controlled release to target tissues [10, 19]. Nanomedicine is an evolving field applying nanoscience expertise and medical biology techniques and the prevention and remediation of diseases. It includes the use of nanodimensional materials, including nanorobots, diagnostics, transmission and sensory nanosensors, and the activation of materials in live cells. For example, a nanoparticle-based approach has been developed that incorporates both treatment and imaging modalities of cancer diagnosis [20]. Lipid systems, including liposomes and micelles, which are now Food and Drug Administration (FDA) approved [21], are used in the first wave of nanoparticle-based therapy. Inorganic nanoparticles such as gold or magnetic nanoparticles may be present in these liposomes and micelles [22]. Such characteristics cause inorganic nanoparticles to increase, emphasizing drug delivery and imaging and therapeutic functions. Nanostructures often allegedly help avoid drugs from becoming tarnished in the gastrointestinal area and help deliver sparingly water-soluble drugs to their target place. Nanodrugs display greater oral bioavailability because they possess traditional absorptive endocytosis mechanisms of absorption.

For an extended period, nanostructures remain in the blood circulatory system and allow the release of composite drugs according to the dose defined. Therefore, more minor plasma variations are induced, with decreased adverse effects [23]. When nanosized, these structures penetrate the tissue system, promote fast drug absorption by cells, allow efficient drug delivery, and ensure action at the desired site. The cell absorption of nanostructures is much greater than that of large particles with sizes varying from 1 to 10 μm [17, 24]. Therefore, with increased performance and decreased or negligible side effects, they interact directly to treat the diseased cells.

Because of their use in various novel assays to treat and diagnose diseases, the nanoparticles are useful in acquiring data. These nanoparticles' key advantages are correlated with their surface properties; the surface may be affixed with different proteins. For example, gold nanoparticles are used as biomarkers and tumour labels for various procedural assays for biomolecule detection. Concerning nanomaterials in the delivery of drugs, the option of a nanoparticle depends on the drugs' physicochemical characteristics. In recent times, the combined usage of nanoscience and bioactive natural compounds is attractive and growing. The distribution of natural products for the treatment of cancer and several other diseases has many benefits. Because of their various characteristic functions, such as inducing tumour-suppressing autophagy and functioning as antimicrobial agents, natural compounds have been comprehensively studied in curing diseases. Autophagy was found in curcumin and caffeine [25], while cinnamaldehyde, carvacrol, curcumin, and eugenol [26, 27] were shown to have an antimicrobial impact.

The enrichment of their properties, such as bioavailability, targeting, and controlled release, is made possible by integrating nanoparticles. For example, thymoquinone, a bioactive compound in *Nigella sativa*, is studied in the lipid nanocarrier after its encapsulation. Encapsulation has demonstrated a sixfold improvement in bioavailability compared to free thymoquinone and has preserved gastrointestinal forms of substances [28]. The pharmacokinetic properties of the natural product were also improved, resulting in more substantial therapeutic effects.

With the design of target-specific drug delivery systems, metallic, chemical, inorganic, and polymeric nanostructures, including dendrimers, micelles, and liposomes, are frequently considered. In particular, these nanoparticles are tagged with certain drugs with low solubility with less absorption potential [17, 29]. However, depending on the scale, shape, and other inherent biophysical/chemical characteristics, these nanostructures' effectiveness as drug delivery vehicles varies. Polymeric nanomaterials with diameters ranging from 10 to 1000 nm, for example, exhibit outstanding features for an effective delivery vehicle [7]. Various synthetic polymers such as polyvinyl alcohol, poly-l-lactic acid, polyethylene glycol (PEG), and poly(lactic-co-glycolic acid) and natural polymers such as alginate and chitosan are commonly used in the nanofabrication of nanoparticles because of their high biocompatibility and biodegradability properties [8, 30–32]. Polymeric nanoparticles, both excellent drug delivery devices, can be categorized into nanospheres and nanocapsules. Similarly, compact lipid nanostructures and phospholipids, including liposomes and micelles, are very useful in targeted drug delivery.

The use of the optimal delivery mechanism for nanodrugs is determined primarily based on the biophysical and biochemical properties of the treatment-selected targeted drugs [8]. However, when contemplating the use of nanomedicine, issues such as toxicity demonstrated by nanoparticles cannot be overlooked. More recently, in conjunction with natural materials, nanoparticles have predominantly been used to mitigate toxicity problems. As it curtails the biosynthetic process's harmful constituents, the green chemistry route of developing nanoparticles loaded with drugs is widely encouraged. Therefore, using green nanoparticles to deliver drugs will reduce the side effects of the drugs [19].

Besides, adjustments to the size, shape, hydrophobic, and surface changes of nanostructures may further enhance their bioactivities. Thus, through site-specific and target-oriented medicines, nanotechnology provides several benefits in treating chronic human diseases. However, inadequate awareness of nanostructures' toxicity is a significant concern and certainly warrants more research to enhance efficacy with higher protection to allow for safer practical application of these medicines. Carefully designing these nanoparticles could therefore help to address the problems associated with their use. The man focus in this chapter is on different nano-based drug delivery systems, significant applications of natural compound-based nanomedicine, bioavailability, and targeting sites, controlled release of nanodrugs, and other challenges related to nanomaterials in medicinal products.

9.2 Phytochemicals and Nanoparticles

Phytochemicals contain various chemical structures, represented by phenolic acids, indoles, alkaloids, isothiocyanates, phytosterols, saponins, and phytoprostanes/furans [33, 34]. These compounds have received increasing attention, boosted by their human health benefits [35, 36], to many mechanisms, including free radical scavenging, inhibition of microtubule assembly and microfilament, chelation of metals, and inhibition of protease, among others [37]. Therefore, the assessment of these compounds, from their in vivo metabolization point of view, has shown that phytochemicals are recognized and processed as xenobiotics after oral intake and are degraded as a result of various chemical, enzymatic, and microbial digestion phases in the mouth, stomach, and small and large intestines and then absorbed, distributed, and metabolized in cells and tissues. In the meantime, the maximum bioaccessibility and bioavailability of phytochemicals give rise to their actual ability to exert biological effects in vivo (bioefficacy). Bioactive nutrients and non-nutrient are demonstrated by short-term changes in liver function biomarkers' expression, plasma lipid profiles, blood pressure, plasma glucose, and plasma antioxidant activity [38]. Bioefficiency is influenced by the chemical nature of these substances and their metabolic derivatives. Numerous articles published to date on the bioavailability of phytochemicals after oral intake has shown that these compounds or their metabolites may fall within the nano/picomolar range in cells and tissues, which may result in insufficient doses to achieve biological efficacy, have been demonstrated by in vitro characterizations [39].

The dispersion and absorption of such compounds depend closely on their polarity. When ingested, cellular absorption, efflux pumping, and phase II and phase I metabolism are caused by the small intestinal enterocytes. This method results in the sum of bioaccessible phytochemicals [40]. The transformations occurring in the frame of phase I and II reactions in hepatocytes give rise to active setups from inactive precursors that are the actual responses of their biological value [41]. Besides, the large intestine enters unabsorbed phytochemicals where the local microbiota is further metabolized. Their bioefficacy is also influenced by the further modification, which produces new bioactive compounds that lead to a specific plant matrix [40].

Edible nanoencapsulation vehicles have been noted as a valuable alternative to avoid the adverse effects of metabolization upon actual bioactive compounds in cells and tissues. A variety of healthy ingredients are currently being used to produce edible nanoparticles with well-defined sizes, surface properties, matrix materials, etc., including lipid, polysaccharides, proteins, and biologically degradable polymers. These nanoparticles are engineered to provide improved strategies to enhance the release and enhanced bioefficiency of compounds of concern for target cells and tissues [42, 43]. A wide range of structural features are found in various types of nanoparticles suited to the use of bioactive phytochemicals in oral administration, such as nanoliposome, micelle, nanoemulsion, solid lipid nanoparticles (SLNs), polymeric nanoparticles, protein–polysaccharide complex coacervation, and the incorporation of cyclodextrin and polymeric nanogels. Indeed, these alternative nanoparticles have several characteristics, including the loading phase of phytochemical compounds, the efficiency of encapsulation, the stability during gastrointestinal digestion, the releasing mechanisms, and, consequently, their ability to increase the bioefficacy of phytochemicals carried [40]. The recent assessment of neuroblastoma cells' capacity to tolerate nanoparticles with encouraging results has shown this situation. The findings show that nanoparticles lead to adequate amounts of nervous system bioactive compounds that are more capable of the blood–brain barrier than cationic ones to move successfully [44]. Besides, it is possible to achieve additional stability for bioactive compounds in cells and tissues in vivo [45, 46] by using this form of nanoparticles, given the repulsion among the highly negatively charged nanoparticles. These characteristics influence bioefficacy, dispersion and gastrointestinal stability, release rate and delivery location, endothelial cell layer transport effectiveness, systemic distribution, and their ability to regulate the impact of microbiota's metabolism [40]. Besides, it is essential to note that particle size, along with polymeric nanoparticles, SLNs, or nanoemulsions, contributes to stability while improving cellular uptake by affecting repulsion facts interfacial tension decreasing compounds [47, 48], which contribute to increased cellular uptake.

The loading phytochemicals within such edible nanoparticles enhance the dispersion of initially water-insoluble phytochemicals in aqueous media [49]. In contrast, the physical barrier provided by nanoparticles protects bioactive compounds against oxidation under acidic and alkaline degradative conditions in the stomach and small intestine respectively [50, 51]. There have been frequent records of the bioactivity of phytochemicals in their native state. Simultaneously, the distribution of bioactive nutrients and non-nutrient in vivo is critically impaired by their solubility, affecting stability and penetration under gastrointestinal conditions, as described before. Therefore, phytochemicals' encapsulation offers practical alternatives to the release of bioactive substances, affected by pathophysiological conditions, in particular tissues. In this context, bioactive compounds' transport to nanoparticles restricts their degradation and general metabolism under the gastrointestinal tract's chemical and enzymatic conditions. In this regard, the physicochemical characteristics of the materials used for the manufacture of nanoparticles and the digestibility of nanoparticles in the intestinal or gastric parts are vital issues that should be discussed, first of all, in the production of a nanoparticle-based strategy for the oral administration of phytochemicals [40]. Some illustrations like although starch-based nanoparticles are digested by α-amylase activity at the oral stage, other polysaccharide nanoparticles are degraded in the small intestine, and protein–polysaccharide nanoparticles may be attributable to changes in pH and salt concentrations, such as those that occur in the gastrointestinal lumen throughout the separate digestion processes. In the small intestine, lipid nanovesicles release phytochemicals simultaneously to digest triglycerides [40]. The loading method is essential for the final release of phytochemicals and, therefore, for their bioavailability and bioactivity, in addition to the existence of the nanoparticle. In this regard, the loading efficiency is strongly affected by the drug loading method. It has been defined that nanoencapsulated compounds release to a lower extent than nanosphere-based delivery, even though both formulations are of similar efficiency [52, 53]. A new compound, plant oxylipins, such as phytoprostanes and phytofurans formed by non-enzymatic α-linolenic acid oxidation, has recently emerged. Based on their structural analogy to human eicosanoids, bioavailability and biological activity of such compounds have been suggested [54]. As they migrate into fatty acid-based micelles, these lipid phytochemicals are compounds with unique characteristics, leading to their solubility and subsequent cellular transport [40].

In addition to regulating the release of phytochemicals, nanoparticles also influence the transport of bioactive compounds through enterocytes by modulating the time of residence, transport efficiency, pathways, and metabolic reactions within endothelial cells of such compounds. For instance, chitosan derivatives have been linked to mucoadhesive characteristics and their ability to open tight junctions, enhancing paracellular transport [55]. The nanoparticle's charge also affects the formation of hydrogen bonds with the mucosal surface, leading to momentary retention [56]. Incorporating cell-penetrating ligands to improve transmembrane transport efficiency is another feature of nanoparticles that influences their position as carriers of bioactive compounds [40]. Nanoparticles filled with bioactive phytochemicals pass through the endothelial layer to migrate to tissues in the bloodstream until absorbed from the intestinal lumen. At this stage, nanoparticles have already been included in endosomes. They have undergone, among other things, oxidation, reduction, and hydrolysis reactions, as well as conjugations following phase II metabolism by various metabolization routes. However, unlike the metabolism of free phytochemicals, when administered with nanoparticles, bioactive compounds are shielded against these reactions by modulating enterocytes' sensitivity to metabolizing enzymes [57]. Highly lipophilic phytochemicals with unique characteristics, such as the oil–water partition coefficient logarithm and long-chain solubility of triglycerides (e.g. phytoprostanes and phytofurans), can move through enterocytes and form enterocyte lipoprotein chylomicrons [58]. Through mesenteric lymph and thoracic ducts, these reach the bloodstream, preventing hepatic first-pass metabolism. Nanoparticles containing a lipid phase (for example, emulsions, liposomes, and SLNs) have been used to imitate this absorption path, enhancing the involvement of lymphatic absorption when the transcellular absorption rate increases [59]. Because of the interest in incorporating the lipid process in nanoparticles as absorption rate modulators, the degree to which the phospholipid composition and emulsifier concentration can fine-tune the biodistribution by altering the lipophilicity and plasma binding properties of phytochemicals has been investigated [60]. In comparison to approaches that have been explored, the use of nanoparticles as carriers as a practical means of monitoring the release and bioavailability of bioactive nutrients and non-nutrients administered, the use of magnetic nanoparticles, the modification of surface chemical properties, and the inclusion of ligands that promote cellular uptake are represented [40]. However, some types of compounds, such as stilbenes, flavonols, and anthocyanins, which are metabolized by the intestinal microbiota towards even more powerful bioactive compounds, could be interesting avoiding their digestion and absorption at gastric and small intestine levels. With this objective, the best option lies in using nanoparticles manufactured with materials or multilayer structures resistant to gastric and small intestine conditions in terms of pH, osmotic conditions, and enzymatic activities. It is also essential to design release mechanisms that enable the delivery of bioactive phytochemicals under the large intestine's osmotic and pH conditions to accomplish this task and microbiota responsible for its transformation [40].

On the other hand, after this goal has been accomplished, the difficulty of obtaining highly bioavailable compounds extracted from microbiota metabolism constitutes an additional restriction, owing to low colonic absorption at luminal fluid volume viscosity and neutral colon pH conditions [61]. Once consumed and circulated in the bloodstream, various methods, such as hydrophilic coating, have been proposed and explored to regulate circulation duration. By acting as a steric barrier and thus shielding nanoparticles from opsonization, PEG is reliable for improving circulation time [62]. Poly(lactic-co-glycolic acid) is one of the most frequently used products for forming nanoparticles, representing a successfully used biodegradable polymer because of its physiological hydrolysis of the metabolite monomers' lactic acid output glycolic acid, which is rapidly processed via the Krebs cycle. Low systemic toxicity has been described as using this biomaterial to regulate the bioactive compound's release because of this rapid metabolization [63]. However, it is necessary to develop surface modifications using non-toxic and blood-compatible materials that allow macrophages to absorb them and increase the length of blood circulation, thereby contributing to the sustained delivery of bioactive compounds carried in this nanosystem [64]. Poly(ethylene glycol) is used in this relation as a non-toxic hydrophilic segment combined with hydrophobic biodegradable aliphatic polyesters.

9.3 Bioactive Delivery Systems

Numerous therapeutic procedures have been proposed to improve drug specificity and diagnostic precision with nanomedicine's advancement and the development of drug discovery and design or drug delivery systems. Therefore, the critical goal is to reduce their toxicity and increase their bioavailability in the body [65].

A promising feature has been the drug design that characterizes novel lead drugs based on a biological target's knowledge. Progress in computer science and the advancement of experimental procedures to categorize and purify proteins, peptides, and biological targets are essential for this sector's growth and development [66, 67]. Besides, several studies and reviews have been discovered in this field, focusing on the rational design of different molecules and demonstrating the importance of studying drug release mechanisms [68]. Besides, natural products can provide viable and exciting solutions to drug design challenges and serve as an inspiration for discovering drugs with desired physicochemical properties [69, 70]. In the past few years, drug delivery technologies have also been gaining prominence. Such systems can be rapidly developed and can facilitate the modified release of the body's AIs. Chen et al. [70], for instance, described a fascinating study of nanocarriers for imaging and sensory applications and addressed these systems' therapeutic impact. Besides, Pelaz et al. [71] presented an up-to-date summary of a range of nanocarriers' applications for nanomedicine and addressed emerging opportunities and challenges for the field. Interestingly, each drug delivery system has its chemical, physical, and morphological characteristics, and through chemical interactions or physical interactions, it may have an affinity for different drug polarities. Mattos et al. [72] showed that the release profile of biogenic silica nanoparticles grafted with neem bark extract was lower than biogenic silica nanoparticles loaded with neem bark extract. These factors affect nanocarriers' interaction with biological systems [73] and the release kinetics of the active material in the organism [68]. Besides, a cross-linkable lipid shell (CLS) containing docetaxel and wortmannin was designed by Sethi et al. [74] as the prototypical drugs used to monitor the kinetics of drug discharge; their discharge profile was then tested, which was found to be affected in both in vivo and in vitro conditions. Understanding their drug delivery profile, nanocarriers' composition, and how drugs are associated with them is also essential [75, 76]. A few mechanisms reflecting the release of drugs in nanocarriers is diffusion, solvent, chemical reaction, and stimuli-controlled release, as shown in Figure 9.1 [77, 78].

Kamaly et al. [79] presented a wide range of analysis of controlled release systems based on studies related to polymeric nanocarrier drug release regulation. Although there are many nanocarriers with different drug release profiles, strategies are currently being developed to boost the specificity of nanostructures in the target regions of the organism [80] and to minimize immunogenicity by coating them or by chemical fictionalization with several

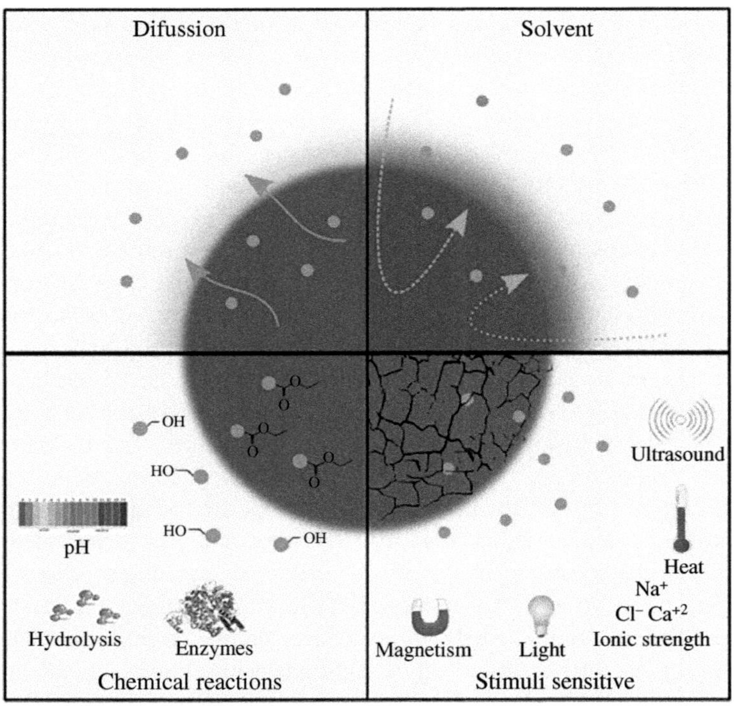

Figure 9.1 Pathway mechanisms for the controlled release of drugs using nanocarriers. Source: Patra et al. [77].

substances, such as polymers [81], natural polysaccharides [82, 83], antibodies [84], cell memory [83], and antibodies [84]. These ligand-modified nanocarriers have been used in some instances where drugs do not show binding and affinity with a specific target or do not cross those barriers [85] to move through the cell membrane-programmed drug delivery in a particular environment. For example, hyaluronic acid has been used in several nanocarriers as a ligand showing promising results to improve antitumour activity against melanoma stem-like cells [86], breast cancer cells [53], pulmonary adenocarcinoma cells [87], and to promote the delivery of intravitreal drugs for retinal gene therapy [83] and to decrease retinal gene therapy [83]. The architecture of the ligand-appended drug delivery systems, however, is labour-intensive. Considering the physiological variables of blood flow, disease status, and tissue architecture [88], multiple targeting designs must be carried out beforehand.

Besides, few studies have been conducted to test ligand-appended interaction with cell membranes in nanocarriers, and their uptake mechanism is still unclear. It has also been recognized that the absorption of nanoparticles by cells occurs through phagocytic or non-phagocytic pathways [89, 90]. However, it has been difficult to standardize the mechanism of action/interaction of these systems in cells because of each delivery system's specific essential physicochemical characteristics. For example, the critical endocytosis mechanisms responsible for cellular uptake of polysaccharide nanoparticles containing active compounds were highlighted by Salatin and Yari Khosroushahi [91].

On the other hand, the ability to monitor the release profile of drugs using external factors such as ultrasound [92], heat [93–95], magnetism [96, 97], light [98], pH [99], and ionic strength [100] has been shown by stimulus-responsive nanocarriers, which can enhance targeting and allow greater dosage control. For example, polymeric nanocarriers [101] or lipids [102] are associated with superparamagnetic iron oxide nanoparticles to trigger a controlled release mechanism by applying an external magnetic field. Ulbrich et al. [103] explored drug delivery mechanisms primarily based on polymeric and magnetic nanoparticles, as well as the effects of covalently or non-covalently attached drugs on the treatment of cancer [103]. Currently, hybrid nanocarriers are among the most promising methods for nanomedicine. A single system presents a mixture of different systems' properties, thereby ensuring products with better efficiency for both therapeutic and diagnostic applications. Notwithstanding this, little is understood about the basic mechanisms of action and toxicity of drug delivery systems, opening the door to recent studies [10].

9.3.1 Nanotechnology of Natural Products and Drug Delivery

In developing countries, about 80% of the population's primary health needs are fulfilled or complemented by conventional medicine [104]. The scientific community focuses on research on bioactive substances, their chemical structure, and the pharmacological ability of different plant species to develop novel AIs with relatively minor side effects compared to existing molecules [105]. Plants have long been recorded as significant sources of natural compounds of medicinal significance. Nevertheless, it has enough resources for the discovery of new and extremely efficient medicines. Nevertheless, some problems are associated with the detection of active compounds from natural sources because they come from living beings whose metabolite composition changes in the presence of stress. The pharmaceutical companies have agreed to merge their efforts to produce synthetic compounds in this sense [106].

Nevertheless, day-by-day, the number of synthetic molecules being sold will decrease. Thus, research on natural product-based active compounds is once again coming to the fore despite its challenges. Almost all of the natural economic value substances with therapeutic potential that have already been marketed have been found in higher plants [107]. Several medications are now commercially available that also have natural therapeutic agents in their composition; their applications and names are as follows: malaria treatment (Artemotil derived from *Artemisia annua* L), Alzheimer's disease treatment (Reminyl – an acetylcholinesterase inhibitor isolated from the *Galanthus woronowii* Losinsk), cancer treatment (Paclitaxel® – *Taxus brevifolia* plant; vinblastine and vincristine extracted from *Catharanthus roseus*; camptothecin and its analogues derived from *Camptotheca acuminate* Decne), and liver disease treatment (silymarin – *Silybum marianum*) [105].

It has already studied and developed the structure and behaviour of many natural compounds. The bioactive molecules present in plants are alkaloids, flavonoids, tannins, terpenes, saponins, steroids, and phenolic compounds. Even so, in most cases, because of the absence of the ability to cross the lipid membranes due to their large molecular sizes, these compounds have poor absorption potential, resulting in a decrease in bioavailability and efficacy [108]. These molecules also demonstrate high systemic clearance, which needs repeated applications and high doses, decreasing the medication's potency for clinical use [106]. The scientific advancement of nanotechnology will revolutionize the development of natural product-based formulations, bringing tools capable of solving the problems mentioned above that restrict the large-scale application of these compounds in nanomedicine [106, 109]. Nanotechnology techniques in the medical field have been extensively studied in the past few years [110, 111] and overcome these barriers and allow the preparation of the same formulation for various compounds and mixtures. Also, inside the biological system, they can alter properties and behaviour [106]. Also, release systems guide the compound to the specific site, increase bioavailability, expand compound action, and combine molecules with varying hydrophilicity/lipophilicity, providing benefits to the compound compared to the solubility and stability of the compounds. The association of release systems with natural compounds can help delay the development of drug resistance and, thus, play an essential role in identifying new possibilities for treating several diseases that do not react adequately to conventional approaches to modern medicine [106, 109]. There are two types of natural product-based materials that are aimed at a specific location and released at specific sites for the treatment of many diseases [112] and are mainly used in the synthesis process [113]. Most of the study is intended for cancer disease care because it is the world's leading cause of death today [114, 115]. In the case of cancer, it affects various organs of the body. Therefore, the creation of alternative medicines to target cancer cells is the highest priority among modern researchers. However, there is also work on nanomedicine applications for other ailments [116]. These delivery systems are classified in terms of their surface charge, particle scale, dispersion size, shape, stability, capacity for encapsulation, and biological action, which are further used according to their requirements [33]. On an ongoing basis, the pharmaceutical industry has tried to develop and apply new technologies to advance and construct modern drugs and to reinforce the existing ones [117]. Via numerous techniques, the rapid progress of nanotechnology has guided the development of new formulations, such as driving the drug to the site of action (nanopharmaceuticals); image and diagnosis (nanodiagnosis), medical implants (nanobiomaterials), and the combination of disease diagnosis and treatment (nanotherapy) [118, 119].

Several nanomedicines currently under development are updated release systems for active ingredients (AIs) that are already used in patients' care [120]. In this context, it can be ascertained that the nano-formulation offers advantages over the existing formulation if the AI is directed towards the target tissue shows increased uptake/absorption by the cells and lower toxicity profile for the organism [121]. Berberine, curcumin, ellagic acid, resveratrol, curcumin, and quercetin are covered in this portion. Doxorubicin (DOX), paclitaxel, and vancomycin are some of the other compounds listed that also come from natural products. Using natural materials, nanoparticles were synthesized. For instance, various microorganisms, including bacteria, fungi, algae, yeast, and so on [122], or plant extracts [123] have been used to synthesize metallic, metal oxide, and sulfide nanoparticles. The microorganism that supports the synthesis procedure is prepared in the required growth medium and then mixed in solution with a metal precursor and left to form intracellular or extracellular nanoparticles for incubation [124]. Besides, the plant extract is prepared and subsequently mixed in solution with the metal precursor and further incubated for a definite duration at room temperature or boiling temperature or exposed to light as an external stimulus to initiate the synthesis of nanoparticles [125].

These natural product-based materials are currently considered essential ingredients in preparing and processing new nanoformulations because they have exciting features such as biodegradability, biocompatibility, availability and renewability, and low toxicity [126]. Biomaterials are, for the most part, capable of undergoing chemical modifications in addition to the properties mentioned above, ensuring unique and attractive properties for future uses in the field of nanomedicine [127, 128]. Several bacteria, namely, *Escherichia coli*, *Pseudomonas aeruginosa*, *Bacillus subtilis*, and *Klebsiella pneumonia* [124], have been used to synthesize gold, silver, cadmium sulfide, and titanium dioxide with different morphological characteristics. Because of their higher potential among all metal nanoparticles, these nanoparticles, especially silver nanoparticles, have been extensively studied in vitro for their antibacterial, antifungal, and cytotoxicity potential [129]. Maximum research focuses on how microorganisms lower metal precursors and manufacture the nanoparticles in the case of microorganism-mediated nanoparticle synthesis. For example, Rahimi et al. [130] used *Candida albicans* to synthesize silver nanoparticles and investigate their antibacterial activity against two pathogenic bacteria, namely, *Staphylococcus aureus* and *E. coli*.

Similarly, Ali et al. [131] synthesized silver nanoparticles with the *Artemisia absinthium* aqueous extract, and their antimicrobial activity was assessed versus *Phytophthora parasitica* and *Phytophthora capsici* [131]. Malapermal et al. [132] used *Ocimum basilicum* and *Ocimum sanctum* extracts to synthesize nanoparticles and studied its antimicrobial potential against *E. coli*, *Salmonella* spp., *S. aureus*, and *P. aeruginosa* along with the antidiabetic potential. Sankar et al. [133] also tested silver nanoparticles' impact against the human lung cancer cell line for antibacterial and anticancer potential. In addition to the use of microorganisms, they synthesized nanoparticles of silver, gold, and iron oxide using various food waste materials, such as extracts of *Zea mays* leaves [134, 135], onion peel extract [136], *Z. mays* silk hair [137], *Cucumis melo* and *Prunus persica* fruit outer peel

[138], *P. persica* outer peel [139], watermelon rind extract [140], etc. Potential antibacterial activities against different foodborne pathogenic bacteria anticandidal activity against many pathogenic *Candida* spp., potential antioxidant activity and inhibitory proteasome effects have been studied. Crystal nanoparticles, liposomes, micelles, polymeric nanoparticles, SLNs, superparamagnetic iron oxide nanoparticles, and dendrimers are the most widely studied nanocarriers for drug delivery purposes [141]. For natural product-based drug delivery, all of these nanocarriers are formulated. Gupta et al. [142] synthesized chitosan-based nanoparticles loaded with paclitaxel (Taxol) derived from *T. brevifolia* and used to treat various forms of cancer for cancer treatment applications. The authors concluded that with sustained release, high cell uptake, and decreased haemolytic toxicity, the nanoparticle-loaded drug showed improved activity compared to pure paclitaxel [142, 143]. Berberine is a barberry alkaloid herb. To increase suppressive *Helicobacter pylori* growth and to minimize cytotoxic effects in infected cells, Chang et al. [144] developed a heparin/berberine conjugate [144]. To encapsulate vancomycin, Aldawsari and Hosny [145] synthesized ellagic acid SLNs (a glycopeptide antibiotic produced in the cultures of *Amycolatopsis orientalis*). Besides, in vivo rabbit studies have shown that ellagic acid has prevented the development of free oxygen radicals and their clearance radicals, thus preventing damage and encouraging repair [145]. Quercetin has antioxidant properties and is a polyphenol that belongs to the flavonoid group and can be found in citrus fruits and vegetables. In a study conducted by Dian et al. [146] polymeric micelles were used to deliver quercetin, and the results showed that these micelles were capable of delivering continuously in vitro release for up to 10 days, with continuous plasma levels and improved full in vivo drug accessibility [146]. Daunorubicin is a natural product derived from various wild strains of *Streptomyces* and the hydroxylated form of DOX used in chemotherapy [126]. The multifunctional liquid crystal nanoparticle system for intracellular fluorescent imaging and DOX delivery was developed by Spillmann et al. [147], in which the nanoparticles were functionalized with transferrin. Cellular uptake and sustained release were achieved in HEK 293T/17 cells inside endocytic vesicles. Perylene was used to monitor the particles and encapsulate intracellular delivery agents as a chromophore [147, 148]. Purama et al. [149, 150] extracted dextran from two sucrose-based lactic acid bacteria, namely, *Streptococcus mutans* and *Leuconostoc mesenteroides*. Agarwal et al. [151] formulated a dextran-based dendrimer formulation and evaluated its drug discharge capacity and haemolytic activity under in vitro condition. They concluded that the dendritic structure selectively enters the highly porous portion of the affected cells without disturbing the healthy tissues, thereby making more convenient for its application in the biomedical field [151]. Doxil (a form of DOX that was the first FDA-approved nanodrug in 1995) [152] was also delivered by folate-functionalized superparamagnetic iron oxide nanoparticles previously developed for the cure of liver cancer. The in vivo experiments in rabbits and rats decreased two and four times relative to Doxil alone, while folate helped and improved precise targeting [153]. Liposomes are the nanostructures that have been most studied and have been used to deliver natural products such as resveratrol [154] in many formulations. Curcumin, a turmeric-derived polyphenolic compound, has been reported to be used to treat cancers, including breast, bone, cervix, liver, lung, and prostate cancer [155]. Formulations of liposomal curcumin have been developed for cancer treatment [156]. Cheng et al. [157] used various methods to encapsulate curcumin in

liposomes and compared the results resulting from pH dependence that produced stable products with good encapsulation efficiency and bioaccessibility for potential cancer treatment applications [157].

9.3.2 Protein-Based Nanoscale Delivery Systems

Proteins may be the first form of the food-origin biopolymer to be studied in nanotechnology. Proteins are biopolymers of amphiphilic that are biologically built on a mixture of various amino acids. Proteins are non-toxic and extraordinarily biocompatible. They have a strong affinity for bioactive compounds by various molecular interactions, such as hydrogen bonds, hydrophobic and π–π interactions, and disulfide bonds. For encapsulation and delivery purposes, both animal and plant proteins have been extensively studied. The caseins are amphiphilic proteins which, when dispersed in an aqueous process, such as milk, self-assemble into a stable micellar structure. As early as 1968, the micellar structure of caseins was identified and visualized [158], followed by a few different structure models proposed by milk scientists [159]. In 2017, Huppertz et al. proposed the fresh casein micellar structure model. They stated that casein micelles were a porous network of CaP nanocluster-linked non-spherical particles [160]. Despite different structural models, it is widely accepted that casein micelles have a self-assembled nanostructure consisting of four subfractions of phosphopeptides (alps1-, alps2-, β-, and x-caseins) bridged by associative interactions by calcium phosphate nanoclusters. While spherical aggregates with a wide range of sizes (50–500 nm) co-exist in milk, casein micelles' dimension is around 150 nm on average [161]. As a natural delivery vehicle for nutrients in milk, casein micelles serve.

To reconstitute casein micelles in vitro from rehydrated commercial sodium caseinate with a combination of salts (imitating calcium phosphate nanoclusters) with a uniform dimension of 125 nm [162], food scientists have developed (ultra) high-pressure homogenization technology. This technological advance became a landmark in using reconstituted casein micelles to encapsulate different bioactive compounds for food applications. Via hydrophobic interactions during reassembly by simply mixing ethanol-dissolved target compound with caseinate aqueous solution, lipophilic bioactive compounds can be easily trapped in the centre of casein micelles. Relatively recent developments in casein nanoencapsulation technology are manipulating their supramolecular structure at defined conditions to cause micelle self-assembly dissociation and subsequent reassembly upon reversing its original condition (Figure 9.2) [163, 164]. By co-dissolution with dissociated casein micelles and subsequent reassembly to cause strong hydrophobic interactions with a micellar centre, encapsulation of lipophilic molecules is achieved. Calcium-chelating agents such as citrate, alkaline pH, and warm aqueous ethanol solution are conditions that can cause the dissociation of casein micelles. The use of organic solvents and advanced equipment is eliminated by this modern nanoencapsulation method. Significantly, among the three methods, the pH-driven method, i.e. increasing pH to alkaline conditions followed by neutralization to pH 7, is the easiest, simplest, and most economical method. In many other colloidal structures, such as protein nanoparticles

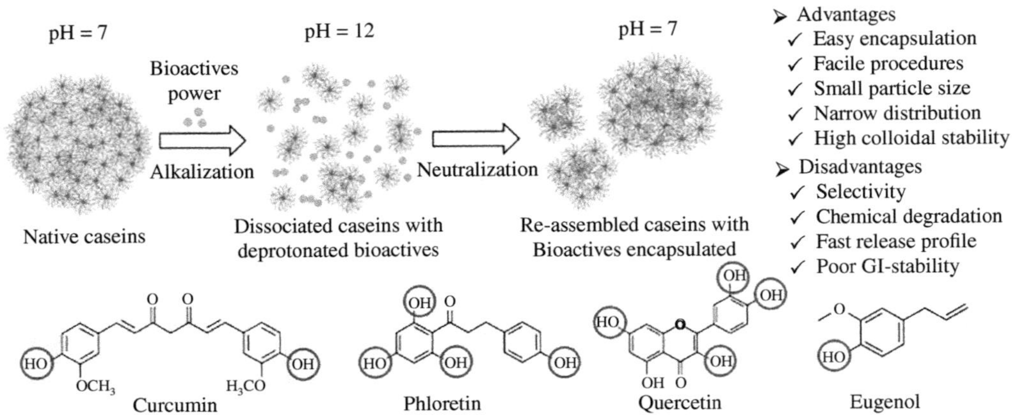

Figure 9.2 The Representation of pH-driven nanoencapsulation of phytochemicals into casein nanoparticles. The small circles in chemical structures indicate the deprotonated hydroxyl groups at alkaline condition. Source: Luo et al. [163].

[165], protein/polysaccharide complex nanoparticles [166, 167], and lipid-based systems [168], this approach has been exploited. Nevertheless, for co-dissolution with dissociated casein micelles, this approach is selective in encapsulating the lipophilic bioactive with a phenolic group deprotonated under alkaline conditions. Besides, a recent study shows that chemical caseinate modification with selected alkyl groups is another technique to induce caseinate micelles to be reassembled, which significantly enhances their self-assembly properties and loading potential for bioactive lipophilics [169].

Another food-derived protein commonly studied in nanotechnology is zein, prolamine maize [170]. Zein has unique solubility in 50–95% aqueous alcohol solutions. As a result of its specialized tertiary structure with an elongated prism-like shape of nanoscale dimensions [171], zein molecules exist as nanoaggregates instead of monomers an aqueous alcohol solution. Provided this attribute, zein nanoparticles can be readily prepared by either liquid–liquid dispersion [172] or a more recently established solvent–evaporation-induced self-assembly process [173]. The former method is based on the rapid reduction of its solubility to induce the formation and solidification of zein nanoparticles. The magnitude depends on the shear force and the final concentration of alcohol. This latter technique is based on the preferential evaporation of one of the solvents that alter the zein solution's polarity, which drives its self-assembly. Zein can self-assemble into various nanoscale structures, including spheres and hexagonal, sponge, and lamellar phases, depending on the polarity variation of binary or tertiary solvents. Because of its easy encapsulation of lipophilic molecules through strong hydrophobic interactions, liquid–liquid dispersion is more applicable to the delivery of bioactive compounds. With a suitable aqueous stabilizer, such as a small molecular surfactant or macromolecular hydrocolloids, zein nanoparticles' colloidal stability can be significantly improved [174, 175]. A recently developed approach to using sodium bicarbonate as a sacrificial template in the preparation of a hollow nanostructure compared to zein nanoparticles with a solid core has significantly improved loading capability and sustained release profile [176, 177]. Another technique to boost its versatility as an oral delivery vehicle is the cross-linking of zein nanoparticles. In particular, the protection and compatibility of food applications are guaranteed by using polycarboxylic acids (such as citric acid) as non-toxic chemical cross-linkers. Zein nanoparticles had a negligible effect of increasing the oral bioavailability of daidzin and isoflavone by 2.6-fold relative to daidzin solution [178] without cross-linking. Although in a recent study, when testing cross-linked hollow zein nanoparticles by intravenous injection, a much better distribution efficacy was reported [179]. It would be worth investigating the biological efficacy of these nanoparticles by oral administration soon.

In addition to casein micelles and zein nanoparticles, several other proteins have been exploited for encapsulation and delivery purposes to prepare nanoparticles. Soy protein is well known for its balanced composition of amino acids that are non-polar, polar, and charged. Thus, through various functional groups, it may interact with different bioactive compounds. The two widely used approaches for preparing soy protein isolate (SPI) nanoparticles are cold gelation and desolvation. The former approach is based on alkaline pretreated SPI gelation caused by calcium at ambient temperature and heating to induce partial unfolding and disruption of large particles. The latter process, desolvation, is to slowly apply an antisolvent to manipulate SPI water's solubility (such as ethanol). Research indicated that 80% of final ethanol concentration was optimal for growing the particle size to less than 100 nm [180]. However, to stabilize the nanostructure after removing the antisolvent, the desolvated SPI nanoparticles must be chemically cross-linked, which could reflect the possible cytotoxicity of the final product. Various bioactive substances, including curcumin [180, 181], β-carotene [182], vitamins [183], and linoleic acid [184], have been encapsulated into SPI nanoparticles. Recent studies have investigated SPI's ability and other protein nanostructures as a stabilizer for Pickering emulsion preparation [109, 185]. Suppose protein nanostructures are preloaded with bioactive compounds to prepare Pickering emulsions before being used as the stabilizer, in that case, these systems can serve as a delivery vehicle in various lipid-based food products, such as low-fat yoghurt. Similar physicochemical properties can be generated with soy protein, whey protein, and distilled component β-lactoglobulin nanoparticles using similar methods, i.e. approaches to cold gelation and desolvation [186].

Similarly, gliadin shares a similar amino acid composition with zein as a prolamin from wheat. The liquid–liquid dispersion process, as discussed above for zein nanoparticles [187], prepares its nanoparticles. Nevertheless, gliadin nanoparticles' applications as encapsulation and bioactive compound delivery systems are not as commonly studied as zein nanoparticles.

9.3.3 Polysaccharide-Based Nanoscale Delivery Systems

Polysaccharides are typically hydrophilic and constitute a complex class of natural biopolymers consisting of monosaccharide glycoside linkages. Polysaccharides consist of large functional groups with strong molecular interactions with bioactive compounds, including hydroxyl, amino, and carboxylic acids. Polysaccharides have a relatively basic structure, either linear or branched, depending on their origins and sources [188], in comparison to proteins with sophisticated and versatile tertiary and quaternary polymeric networks. It makes the planning of their nanostructures easy. As polyelectrolyte, a polysaccharide carries a surface load, whether positive or negative, and nanostructures using a loaded polysaccharide are often prepared by ionizing gelation by interacting with an oppositely loaded electrolyte. If, on the other hand, a polysaccharide is a non-polyelectrolyte, i.e. zero surface charge, nanomaterial synthesis is usually based on chemical or mechanical processing technology to break down the nanoscale of the bulky structure. A third scenario utilizes the natural nanostructures in the native polysaccharide, a new field of science. In this case, due care must be taken in extracting and purifying their nanostructures.

Chitosan, a deacetylated product from chitin, is the most extensively studied polysaccharide for oral drug delivery as a result of its unique positive surface charge when dissolved and protonated in an acidic aqueous medium and, more importantly, its strong mucoadhesive property, chitosan nanoparticles are generally rendered by ionic gelling, through electrostatic interactions with wit, chitosan a deacetylated product from chitin is the polysaccharide of most excellent research for oral drug delivery [189]. Chitosan can also be prepared with other polyanion macromolecules, a different polysaccharide, a protein/peptide, or a nuclear acid, in nanoparticles by polyelectrolyte complexation. A variety of other nanotechnologies have also been used for preparing chitosan-like nanoparticles without taking advantage of their positive surface charges, including emulsion–solvent delivery, reverse micellization, and desolvation. To enclose the bioactive compound in chitosan nanoparticles for food uses, the convenience of preparing, low cost, and mild conditions, ionic gelation and polyelectrolyte complexation are practical approaches. For example, during the past 10 years, tea polyphenols have been encapsulated into chitosan nanoparticles using ionizing gelating methods to increase oral biological bioavailability and efficacy [190, 191], especially catechins and (−)-epigallocatechin gallot. Alginate is a polysaccharide that consists primarily of α-L-guluronic acid(G) and βD-mannuronic acid (M), which has a linear association with 1,4-glycoside residues, derived from brown algae and bacteria. This unique composition transforms alginates into an 'egg-box' structure through ionic cross-linking with multivalent cations, particularly calcium, with excellent gelation behaviour [192]. Alginate hydrogels and microspheres are prepared for bioactive compounds, either externally or internally, in food or biomedical applications [193, 194]. Alginate nanoparticles for food purposes have been mainly made with two forming mechanisms at a lower biopolymer concentration: (i) ionic complexation with calcium ions at either the aqueous solution or the oil droplet interface and (ii) emulsification of water-in-oil coupled with the external or internal gelation of alginate alginates at the attachment. In contrast with chitosan, producing nanoparticles with small particle shaped monodispersed alginate is harder. The molecular weight, viscosity, and M/G ratio of various alginates differ significantly from various sources. So far, alginate nanoparticles, including resveratrol [195], caffeine [196], and vitamin E have only encapsulated a limited number of bioactive compounds [197]. Pectin, another essential feature of the plant cell wall, has a linear anionic backbone with a high weight and viscosity of the molecules and is a critical polyelectrolyte polysaccharide. These pectins' structural characteristics are not suitable for nanoparticles, while they are highly functional ingredients in foods and frequently used as a gelling agent and thickener. The non-polyelectrolyte polysaccharides, such as cellulose and starch, have become increasingly common over the past several years because of the continuous development of many new processing technologies in nanoscale structures in the twenty-first century. Many cellulose-based nanomaterials, including nanowhiskers, nanofibrillated cellulose, and cellulose nanofibrils or nanofibers, are commonly represented by nanocellulose [198]. Owing to the high energy costs associated with their insulation processes, nanocellulose as a food additive has not received much attention. New processing technologies have recently been developed for the separation, purification, and characterization of nanocellulose in high demand for sustainable materials. New production processes are commonly classified into two approaches, either top-down procedures for wood and agricultural residue isolation by enzymatic/chemical/physical processes or bacterial bottom-up production from glucose. Although their uses have been based mainly on biomedical issues for decades, they are becoming more and more aware as stabilizers and materials for food packaging [199]. Considering the ongoing developments of their use as intelligent and controlled drug delivery [200], we believe that food scientists will be motivated in due course to explore their applications for bioactive compounds as natural nanoscale and food-grade vehicles.

Phytoglycogen is a high-density plant-based carbon dioxide with a nanoscale structure. This is a significant genotype of sweet maize, the maize mutant sugar-1 (sub1) kernel. Chemically, it is a hyperbranched α-D-glucan made from α-1,4- and α-1,6-linked glucose units, very similar to starch, while structurally, it is a monodisperse dendritic biopolymer with a size of 50–70 nm, depending upon the source, extraction, and purification. Because phytoglycogen nanoparticles have been characterized as hard spheres and highly hydrated colloids with exceptional water stability, their specific interest, such as curcumin, 13 antimicrobic peptide nisin [201], lutein [202], and quercetin [143], has also been shown in nanoscale food carriers for the encapsulation of bioactive compounds in recent years. Over the past three to five years, hydrophobic moieties such as octenyl succinate have modified phytoglycogen nanoparticles to form amphiphilic nanoparticles to stabilize Pickering emulsion [203] effectively. No particular applications have nevertheless been proposed or documented so far for those phytoglycogen-stabilized Pickering emulsions.

9.3.4 Complex or Hybrid Nanoscale Delivery Systems

Production of complex or hybrid nanoparticles from two or more biopolymers in food nanotechnology has become an emerging area. Such delivery systems have colloidal characteristics and efficiencies as they incorporate many biomaterial advantages.

Two proteins with complementary amino acid profiles usually prepare complex protein nanoparticles and overcome a specific imperfect protein property by the complex method. Zine–caseinate nanoparticles are the primarily studied protein–protein complex. Zein is suitable for encapsulating lipophilic bioactive compounds, although its water insolubility limits its food usage. Caseinate is an amphiphilic protein and has an outstanding emulsifying function and has been shown to have an excellent stabilizing effect during a liquid–liquid dispersion on the surface of the zein nanoparticles. The original nanoscale dimension is preserved by caseinate-stabilized zein nanoparticles, which have exceptional colloidal stability in a wide range of pH and ionic strength [186]. A new improved method for further enhancing the colloidal characteristics of caseinate dissolving in a heated zein solution and then dispersing in a pH 8 buffer was introduced [204]. Compositional analysis indicated that casein formed part of the zein nanoparticle matrices and was not isolated during storage by increasing ionic strength with this enhanced preparation method. The zein examined the powerful delivery vehicles with enhanced chemical instability, increased cellular uptake, and enhanced bioactivity – caseinate complex nanoparticles [128, 205]. To improve the regulated distribution characteristics and enzyme resistance, polysaccharide was introduced into the protein complex forming complexes that shape complex ternary sections. Casein becomes the standard choice for producing binary or ternary complex nanoparticles for this heating-induced completion method. Because of its versatile conformation, it has exceptional thermal stability without complex secondary structures. To properly enclose and include a wide variety of bioactive compounds [206, 207], alginate, chitosan, pectin, and dextran have all been tested for the outmost cover layer zein–caseinate complex nanoparticles. Recent research has also shown that these complex nanoparticles' biological fate is significantly improved in the caseinate–chitosan–dextran ternary complex of nanoparticles, which improved the antifibrogenic behaviour of astaxanthin in a hepatic fibrogenesis cell model [208].

In addition to the complex nanoparticles of biopolymers, novel nanocarriers have recently been produced for lipid–biopolymers of hybrid nanoparticles. These hybrid systems use solid lipids as their central matrix for improving encapsulation and lasting release. Meanwhile, the composition contains proteins and polysaccharides as emulsifiers/stabilizers and surfaces to enhance colloidal stability during storages and digestion. Like biopolymer ternary nanoparticles, lipid–biopolymer hybrid nanoparticles were substantially more significant in their size than those in the lipid-core non-lipid nanoparticles [209].

This hybrid compound could resolve two historical obstacles for nanoparticles with vital lipid substances, i.e. (i) low gastrointestinal stability and (ii) extreme and irreversible drying aggregation. Protein–polysaccharide matrix pretreatment through non-toxic chemical reactions improved the stabilization and the particle size of the hybrid nanoparticles and, thus, enhanced the bioactivity of encapsulated bioactive compounds [210]. These newly formed hybrid nanoparticles have shown their promising potential for bioactive organic compounds orally. Investigators have studied numerous methods in functional foods and drinks to encapsulate several micronutrients and nutraceuticals. The epigallocatechin gallate (EGCG) and curcumin co-encapsulation systems are manufactured for delivery into hyaluronic acid, fucoidae, and modified gelatine nanoparticles for anticancer agents [211]. This form's biopolymer nanoparticles are mainly combined through electrostatic interactions between oppositely charged protein groups and polysaccharide molecules. Improving the composition of the method could make relatively small nanoparticles containing both nutraceuticals. Cell culture and feeding studies have shown that co-encapsulation of these nutraceutical inputs has increased these cancers' activity.

In vitro and in vivo experiments have also been used to demonstrate enhanced cholesterol reduction effects in supply systems that contain a blend of EGCG-rich green tea extract and curcumin [212]. The green tea extract and curcumin were powdered and stored in a teabag. Both nutraceutical products were then assumed to disperse from the tea bags after filling hot water during the brewing process. The EGCG and curcumin have been found in a hydrophobic protein centre in biopolymer particles [213]. In polysaccharide-coated zein nanoparticles, Quercetagetin and curcumin are co-encapsulated [214] and piperine and curcumins are co-encapsulated in closely related systems [215]. The co-encapsulation of the nutraceutical mixtures' protein nanoparticles has improved their resistance towards degradation in light or heat exposure. Their antioxidant activity was shown to increase with EGCG and quercetin co-encapsulation in liposomes [216]. A medication (DOX) was co-encapsulated with EGCG and quercetin in pharmaceutical studies [217]. This is a rapidly developing field, and colloidal delivery systems that have been explicitly developed will probably be used more frequently.

The regulated supply was established as a critical step for administering bioactive compounds to decide the timing, tissue/cell, and pathophysiological conditions for release of therapeutic agents. This aim would allow higher local concentrations close to the seen in vitro operational levels to be achieved, thus reducing the total dose administered. The diversity of internal and external factors allows for various factors, including pH, local behaviour, temperature, ultrasound, magnetic field, and light incidence, to monitor the precise release of bioactive phytochemicals [218]. The nanoencapsulation of compounds recognized as potential therapeutic agents has become increasingly attractive as the spectrum of biomaterials in this area has been expanded [219] with practical applications. Polymeric nanocarriers can increase bioavailability, enhance solubility, and extend the shelf life of potentially tricky compounds. For example, oil-cored nanocapsules initially boost bioactive compound administration, enabling targeted distribution and long-term managed to release to reduce dosage and administration frequency, resulting in improved patient compliance [218]. This has led to a promising way for obtaining biomaterials with tailored proprieties, [220] and, as such, exciting applications as stimulant nanocarriers for controlled release of plant chemicals that also provide transport capability through a variety of biological barriers, [221] being explored as a layer-by-layer autoassembly of pH-sensitive building blocks. The use of suitable biocompatible pH-dependent polyelectrolyte for the formation of nanocapsules in the shell can also provide the tools necessary to create shell-permeable nanocarriers [218].

When nanoparticles are filled with bioactive phytochemicals, they become more efficient and lead to site-specific delivery. Once in cells, several nanoparticles slowly release the bioactive phytochemicals that contribute to a sustained therapeutic impact, a further benefit [143]. The use of rhodamine-loaded PEG nanoparticles has recently demonstrated this ability. After this characterization, red fluorescence is visualized with the cytoplasm, which indicates that nanoparticles are internalized by endocytose [222], saving the phospholipidic barrier represented by the cell membrane in hydrophilic compounds [223]. This issue's significant results indicate that carrying nanoparticular bioactive compounds may promote targeted delivery in specific brain areas, reducing biodistribution to the systemic circulation and thus toxic adverse effects. This method offers significant neuroregeneration advantages [224]. To preserve continuous release of systems of neurotrophic proteins, polymer microspheres, scaffolds, and pipes have been used for oral administration and expanded neural regeneration about the implantation of polymers using invasive surgical techniques into the central nervous system [224]. Indeed, this situation is illustrated in the way that biodegradable poly-(lactic-co-glycolic acid) microspheres or in poly-micropollutants are produced by the slow release of the neurotrophic factor from glial cell lines [223].

Although the fundamental importance and benefits of sudden releases of bioactive compounds have not been completely overlooked in recent years with controlled delivery systems, the phenomenon has not been clarified thoroughly in successful plenary theories. Further, the pharmacologically harmful and economically wasteful effects of explosive releases must be considered [150]. An effective formulation and characterization of the new nanoformulations made up of thermosensitive gelling copolymer. When these devices are used, the insignificant amount of bioactive compounds released escape toxicity in target tissue peaks and concentration valleys. The delivery systems produced using this copolymer can thus reduce the side effects of repeated injections to provide therapeutic products that have an interest, without reducing the efficacy of the bioactive compound and serve as a promising platform to prevent pathophysiology complications [225]. However, in vitro, data are used to explain these effects, and assessments in vivo continue to be explained. The protection of such systems and their contribution to the achievement of operational concentrations in target cells and tissues in polymer and lipid-based nanostructured systems was shown by nanoencapsulation and controlled release studies conducted by nanostructured carriers [223]. Because of this and previous studies on the ability of such carriers to protect dysfunctional therapeutic proteins, e.g. against enzymatic degradation and other environmental stressors [224], similar methods may be extended to the administration of bioactive phytochemicals, taking advantage of their biological potential and preventing their in vivo metabolism. In this context, research on the beneficial effects of nanoparticles loaded with anthocyanins against in vitro neurotoxicity has centred on assessing the cytotoxic profile of nanoparticles in the human model of neuroblastoma. Different concentrations of both free anthocyanins and anthocyanins loaded into poly-nanoparticles (lactide-co-clycolide) permitted this study to discard significant cytotoxic effects.

By comparison, the everyday use of nanoparticles loaded with anthocyanins improved the viability of treated cells by shielding them against neurotoxic events [226]. Poly nanoparticles (lactide-co-clycolide) may thus be promising phytochemical carriers with negligible cytotoxicity events. Besides, the nanoparticle, as mentioned above, administration offers the benefit of preventing the removal of an eventual implant positioned as a local source of bioactive compounds [227].

9.4 Toxicity of Biodegradable Nanoparticles

As far as we know, there is no evidence of adverse effects concerning the toxicity associated with the administration of biodegradable nanocarrier systems for bioactive compounds. However, the threats to human health related to long-term usage remain underexplored [63]. Increased development of reactive oxygen species (ROS) and free radicals are correlated with the mechanisms responsible for the harmful effects of nanoparticles in the context of complex biological systems, which could lead to oxidative stress, inflammation, and, subsequently, many crippling pathophysiological situations associated with solvent residues and toxicity of polymers [228]. To shed some light on nanoparticles' actual toxicity, it is required to test the separate trials separately, paying attention to interactions between nanocarriers and biological systems. This way of processing the available toxicity information associated with nanoparticles is essential for establishing the delivery systems' reliability. It is essential to develop their safety profiles, including evaluating immunotoxicity, to successfully translate nanoformulations from the functional level to their practical clinical application. Therefore, some physiological, physicochemical, and molecular aspects include toxicological facts associated with nanocarrier systems.

Despite the apparent interest in understanding these delivery systems' toxicity, however, it has not been extensively studied to date, likely because the industrial application of this technology is in its infancy. From the information available in the literature on this, it is noted that nanocarrier-based toxicological facts are closely related to their size and shape, their biomaterials, and their ability to cross biological barriers [196, 229]. The development of pro-oxidants after administration has been observed about the molecular mechanisms behind the toxicological effects associated with the treatment of bioactive compounds carried on biodegradable nanoparticles, thus disrupting the balance between their output and the cell's detoxification ability. This reality also includes increased inflammatory reactions because of the level of transcription factors that are redox sensitive [230].

The interest in biodegradable nanoparticles and the constraints associated with their toxicity following oral ingestion is associated with the incidence of adverse effects on the liver, kidney, and spleen about nanoparticles' use to bear bioactive compounds addressed to be consumed in foods, water, or drug delivery devices. Besides, even if the mechanisms responsible for triggering the nanoparticle-induced immune response (particularly for non-protein-based carriers) are not specific, there is a growing emphasis on possible allergic reactions [231]. More studies of the immunogenicity of nanocarriers are needed to gain further insights into this problem to understand the conditions under which they are detected by the immune system that merit the generation of specific immune responses [231]. To cause non-IgE-mediated signs of hypersensitivity [232], PEG-grafted liposome infusion has been defined. On the other hand, advantages can be obtained from the immunogenic characteristics of nanoparticles, as these features suggest them as valuable adjuvants, for instance, for the development of vaccines.

9.5 Future Perspectives

Research developments in applying edible nanoparticles to bioactive phytochemicals would be closely related to new development methods that incorporate several structural designs that are more relevant to different types of bioactive compounds. This strategy will help optimize the benefits of loading bioactive compounds on nanoparticles concerning efficient regulation of prolonged release and in situ delivery. The target delivery has not yet been fully developed, which remains a problem with implementation prospects. Materials and manufacturing strategies will allow edible nanoparticles with enhanced properties to be continuously continuous in this sense. Once the specificity and flexibility of available nanoparticles have been improved, oral administration of bioactive compounds using edible nanoparticles would constitute a realistic approach that would allow the actual potential of phytochemicals to prevent and treat specific pathophysiological status because of accurate tissue-targeted delivery. Consideration of intestinal metabolism concerning the actual biological activity of compounds consumed by oral administration requires evaluating the current concept of bioaccessibility and bioavailability, stressing the importance of absorption in the upper gastrointestinal tract. Indeed, for the particular rational design of the nanoparticle method, information on the characteristics of this physiological process, according to the chemical properties of the bioactive compounds of interest, and the derivatives that can be produced during gastrointestinal digestion, is essential.

Another situation to be discussed is how nanoparticles change the pharmacokinetics of bioactive phytochemicals, leading to new applications, such as internalization speed boosters. To determine the pharmacokinetic bioavailability and bioactivity of the bioactive compounds under study, clarification of this situation is necessary for the proper design of the sampling timing. Overall, encapsulated phytochemicals' inappropriate nanoparticles improve bioavailability by shielding them from degradation during storage and gastrointestinal digestion, enhancing solubility in aqueous media, increasing contact time with the intestinal wall, increasing penetration of mucus and intestinal permeation, promoting cellular absorption, and extending residence time inside the intestinal wall. Besides, it should be considered that somehow loading bioactive compounds into nanoparticles may be an exciting technique to prevent their metabolic conversion. However, little is known to date about how cellular signalling pathways vary between phytochemicals linked to free and nanoparticles. In this regard, it is clear that the interaction between phytochemicals and matrix materials, or molecules in complex biological systems (e.g. in mammals), should be further investigated concerning metabolism and bioefficacy [40]. Therefore, as a guide for future assessment of bioefficacy and predictable in vivo properties, the knowledge obtained from studies focusing on the association between the application of bioactive compounds carried by nanoparticles with the metabolic profile, biodistribution, and bioactivity will be helpful. The interactions between phytochemicals and drugs with gut microbiota and the bioefficacy implications of these interactions remain underexplored concerning the final intestinal level, even though they appear to constitute additional challenges. Indeed, extending the information gathered to the use of nanoparticles as carriers of bioactive phytochemicals will be an opportunity to decipher and take advantage of the reciprocal interactions between phytochemicals and microbiota throughout the gut.

Despite the rational use of nanotechnology to improve bioactive compounds' bioavailability and bioactivity, nanoparticles are synthesized by physical and chemical methods through the use of costly and hazardous chemicals, especially metallic biomaterials, which could restrict their actual in vivo application. These constraints are also enclosed to the environmental impact derived of the residues, generated from the nanoparticles synthesis, being required green technologies using no toxic reagents to prepare metal nanoparticles. The synthesis of nanoparticles using environmentally friendly and biocompatible reagents could minimize these processes' side effects in this respect.

References

1 Swamy, M.K. and Sinniah, U.R. (2016). Patchouli (*Pogostemon cablin* Benth.): botany, agrotechnology and biotechnological aspects. *Ind. Crops Prod.* 87: 161–176.

2 Mohanty, S.K., Swamy, M.K., Sinniah, U.R., and Anuradha, M. (2017). *Leptadenia reticulata* (Retz.) Wight & Arn. (Jivanti): botanical, agronomical, phytochemical, pharmacological, and biotechnological aspects. *Molecules* 22: 1019.

3 Rodrigues, T., Reker, D., Schneider, P., and Schneider, G. (2016). Counting on natural products for drug design. *Nat. Chem.* 8: 531.
4 Siddiqui, A.A., Iram, F., Siddiqui, S., and Sahu, K. (2014). Role of natural products in drug discovery process. *Int. J. Drug Dev. Res.* 6 (2): 172–204.
5 Beutler, J.A. (2009). Natural products as a foundation for drug discovery. *Curr. Prot. Pharmacol.* 46 (1): 9–11.
6 Thilakarathna, S.H. and Rupasinghe, H. (2013). Flavonoid bioavailability and attempts for bioavailability enhancement. *Nutrients* 5: 3367–3387.
7 Bonifácio, B.V., da Silva, P.B., dos Santos Ramos, M.A. et al. (2014). Nanotechnology-based drug delivery systems and herbal medicines: a review. *Int. J. Nanomed.* 9: 1.
8 Watkins, R., Wu, L., Zhang, C. et al. (2015). Natural product-based nanomedicine: recent advances and issues. *Int. J. Nanomed.* 10: 6055.
9 Martinho, N., Damgé, C., and Reis, C.P. (2011). Recent advances in drug delivery systems. *J. Biomater. Nanobiotechnol.* 2: 510.
10 Jahangirian, H., Lemraski, E.G., Webster, T.J. et al. (2017). A review of drug delivery systems based on nanotechnology and green chemistry: green nanomedicine. *Int. J. Nanomed.* 12: 2957.
11 Liu, Z., Tabakman, S., Welsher, K., and Dai, H. (2009). Carbon nanotubes in biology and medicine: in vitro and in vivo detection, imaging and drug delivery. *Nano Res.* 2: 85–120.
12 Orive, G., Gascon, A.R., Hernández, R.M. et al. (2004). Techniques: new approaches to the delivery of biopharmaceuticals. *Trends Pharmacol. Sci.* 25: 382–387.
13 Razzacki, S.Z., Thwar, P.K., Yang, M. et al. (2004). Integrated microsystems for controlled drug delivery. *Adv. Drug Delivery Rev.* 56: 185–198.
14 Arayne, M.S., Sultana, N., and Qureshi, F. (2007). Nanoparticles in delivery of cardiovascular drugs. *Pak. J. Pharm. Sci.* 20: 340–348.
15 Patra, J.K. and Baek, K.-H. (2014). Green nanobiotechnology: factors affecting synthesis and characterization techniques. *J. Nanomater.* 2014: 219.
16 Joseph, R.R. and Venkatraman, S.S. (2017). Drug delivery to the eye: what benefits do nanocarriers offer? *Nanomedicine* 12: 683–702.
17 Mirza, A.Z. and Siddiqui, F.A. (2014). Nanomedicine and drug delivery: a mini review. *Int. Nano Lett.* 4: 94.
18 Rudramurthy, G.R., Swamy, M.K., Sinniah, U.R., and Ghasemzadeh, A. (2016). Nanoparticles: alternatives against drug-resistant pathogenic microbes. *Molecules* 21: 836.
19 Lam, P.-L., Wong, W.-Y., Bian, Z. et al. (2017). Recent advances in green nanoparticulate systems for drug delivery: efficient delivery and safety concern. *Nanomedicine* 12: 357–385.
20 Haba, Y., Kojima, C., Harada, A. et al. (2007). Preparation of poly(ethylene glycol)-modified poly(amido amine) dendrimers encapsulating gold nanoparticles and their heat-generating ability. *Langmuir* 23: 5243–5246.
21 Shi, X., Sun, K., and Baker, J.R. Jr. (2008). Spontaneous formation of functionalized dendrimer-stabilized gold nanoparticles. *J. Phys. Chem. C* 112: 8251–8258.

22 Park, S.-H., Oh, S.-G., Mun, J.-Y., and Han, S.-S. (2006). Loading of gold nanoparticles inside the DPPC bilayers of liposome and their effects on membrane fluidities. *Colloids Surf., B* 48: 112–118.

23 de Villiers, M.M., Aramwit, P., and Kwon, G.S. (2008). *Nanotechnology in Drug Delivery*. New York: Springer.

24 Kabanov, A.V., Lemieux, P., Vinogradov, S., and Alakhov, V. (2002). Pluronic® block copolymers: novel functional molecules for gene therapy. *Adv. Drug Delivery Rev.* 54: 223–233.

25 Wang, N. and Feng, Y. (2015). Elaborating the role of natural products-induced autophagy in cancer treatment: achievements and artifacts in the state of the art. *BioMed Res. Int.* 2015: 934207.

26 Ouattara, B., Simard, R.E., Holley, R.A. et al. (1997). Antibacterial activity of selected fatty acids and essential oils against six meat spoilage organisms. *Int. J. Food Microbiol.* 37: 155–162.

27 Sharma, G., Raturi, K., Dang, S. et al. (2014). Combinatorial antimicrobial effect of curcumin with selected phytochemicals on *Staphylococcus epidermidis*. *J. Asian Nat. Prod. Res.* 16: 535–541.

28 Abdelwahab, S.I., Sheikh, B.Y., Taha, M.M.E. et al. (2013). Thymoquinone-loaded nanostructured lipid carriers: preparation, gastroprotection, in vitro toxicity, and pharmacokinetic properties after extravascular administration. *Int. J. Nanomed.* 8: 2163.

29 Krauel, K., Pitaksuteepong, T., Davies, N.M., and Rades, T. (2004). Entrapment of bioactive molecules in poly (alkylcyanoacrylate) nanoparticles. *Am. J. Drug Deliv.* 2: 251–259.

30 Tan, Q., Liu, W., Guo, C., and Zhai, G. (2011). Preparation and evaluation of quercetin-loaded lecithin-chitosan nanoparticles for topical delivery. *Int. J. Nanomed.* 6: 1621.

31 Sanna, V., Roggio, A.M., Siliani, S. et al. (2012). Development of novel cationic chitosan-and anionic alginate–coated poly(D,L-lactide-co-glycolide) nanoparticles for controlled release and light protection of resveratrol. *Int. J. Nanomed.* 7: 5501.

32 Casettari, L. and Illum, L. (2014). Chitosan in nasal delivery systems for therapeutic drugs. *J. Controlled Release* 190: 189–200.

33 Swamy, M.K. and Sinniah, U.R. (2015). A comprehensive review on the phytochemical constituents and pharmacological activities of *Pogostemon cablin* Benth.: an aromatic medicinal plant of industrial importance. *Molecules* 20: 8521–8547.

34 Pinciroli, M., Domínguez-Perles, R., Abellán, A. et al. (2017). Comparative study of the phytoprostane and phytofuran content of *indica* and *japonica* rice (*Oryza sativa* L.) flours. *J. Agric. Food Chem.* 65: 8938–8947.

35 Lu, B.Y., Li, M.Q., and Yin, R. (2016). Phytochemical content, health benefits, and toxicology of common edible flowers: a review (2000–2015). *Crit. Rev. Food Sci. Nutr.* 56: S130–S148.

36 Chang, S.K., Alasalvar, C., and Shahidi, F. (2016). Review of dried fruits: phytochemicals, antioxidant efficacies, and health benefits. *J. Funct. Foods* 21: 113–132.

37 Son, Y.R., Choi, E.H., Kim, G.T. et al. (2016). Bioefficacy of Graviola leaf extracts in scavenging free radicals and upregulating antioxidant genes. *Food Funct.* 7: 861–871.

38 Medina, S., Domínguez-Perles, R., Gil, J.I. et al. (2014). Metabolomics and the diagnosis of human diseases – a guide to the markers and pathophysiological pathways affected. *Curr. Med. Chem.* 21: 823–848.

39 Ting, Y., Jiang, Y., Ho, C.T., and Huang, Q. (2014). Common delivery systems for enhancing in vivo bioavailability and biological efficacy of nutraceuticals. *J. Funct. Foods* 7: 112–128.

40 Xiao, J., Cao, Y., and Huang, Q. (2017). Edible nanoencapsulation vehicles for oral delivery of phytochemicals: a perspective paper. *J. Agric. Food Chem.* 65: 6727–6735.

41 Melo-Filho, C.C., Braga, R.C., and Andrade, C.H. (2014). Advances in methods for predicting phase I metabolism of polyphenols. *Curr. Drug Metab.* 15: 120–126.

42 Huang, Q.R., Yu, H.L., and Ru, Q.M. (2010). Bioavailability and delivery of nutraceuticals using nanotechnology. *J. Food Sci.* 75: R50–R57.

43 Radhakrishnan, R., Kulhari, H., Pooja, D. et al. (2016). Encapsulation of biophenolic phytochemical EGCG within lipid nanoparticles enhances its stability and cytotoxicity against cancer. *Chem. Phys. Lipids* 198: 51–60.

44 Lockman, P.R., Koziara, J.M., Mumper, R.J., and Allen, D.D. (2004). Nanoparticle surface charges alter blood-brain barrier integrity and permeability. *J. Drug Targeting* 12: 635–641.

45 Liu, Y., Li, K., Liu, B., and Feng, S.S. (2010). A strategy for precision engineering of nanoparticles of biodegradable copolymers for quantitative control of targeted drug delivery. *Biomaterials* 31: 9145–9155.

46 Kumari, A., Singla, R., Guliani, A., and Yadav, S.K. (2014). Nanoencapsulation for drug delivery. *EXCLI J.* 13: 265–286.

47 Xiao, J., Nian, S., and Huang, Q.R. (2015). Assembly of kafirin/carboxymethyl chitosan nanoparticles to enhance the cellular uptake of curcumin. *Food Hydrocolloids* 51: 166–175.

48 Kumar, D.V., Verma, P.R.P., and Singh, S.K. (2015). Development and evaluation of biodegradable polymeric nanoparticles for the effective delivery of quercetin using a quality by design approach. *LWT Food Sci. Technol.* 61: 330–338.

49 Pan, K., Zhong, Q.X., and Baek, S.J. (2013). Enhanced dispersibility and bioactivity of curcumin by encapsulation in casein nanocapsules. *J. Agric. Food Chem.* 61: 6036–6043.

50 Hu, B., Ting, Y., Yang, X. et al. (2012). Nanochemoprevention by encapsulation of (−)-epigallocatechin-3-gallate with bioactive peptides/chitosannanoparticles for enhancement of its bioavailability. *Chem. Commun.* 48: 2421–2423.

51 Zou, L.Q., Zheng, B.J., Zhang, R.J. et al. (2016). Food-grade nanoparticles for encapsulation, protection and delivery of curcumin: comparison of lipid, protein, and phospholipid nanoparticles under simulated gastrointestinal conditions. *RSC Adv.* 6: 3126–3136.

52 Ezhilarasi, P.N., Karthik, P., Chhanwal, N., and Anandharamakrishnan, C. (2013). Nanoencapsulation techniques for food bioactive components: a review. *Food Bioprocess Technol.* 6: 628–647.

53 Gao, X., Zhang, J., Xu, Q. et al. (2017). Hyaluronic acid-coated cationic nanostructured lipid carriers for oral vincristine sulfate delivery. *Drug Dev. Ind. Pharm.* 43: 661–667.

54 Domínguez-Perles, R., Abellán, A., León, D. et al. (2018). Sorting out the phytoprostane and phytofuran profile in vegetable oils. *Food Res. Int.* 107: 619–628.

55 Perez, Y.A., Urista, C.M., Martinez, J.I. et al. (2016). Functionalized polymers for enhance oral bioavailability of sensitive molecules. *Polymers* 8: 214.

56 Barua, S. and Mitragotri, S. (2014). Challenges associated with penetration of nanoparticles across cell and tissue barriers: a review of current status and future prospects. *Nano Today* 9: 223–243.

57 Johnson, B.M., Charman, W.N., and Porter, C.J.H. (2001). The impact of P-glycoprotein efflux on enterocyte residence time and enterocyte-based metabolism of verapamil. *J. Pharm. Pharmacol.* 53: 1611–1619.

58 Charman, W.N.A., Noguchi, T., and Stella, V.J. (1986). An experimental system designed to study the in situ intestinal lymphatic transport of lipophilic drugs in anesthetized rats. *Int. J. Pharm.* 33: 155–164.

59 Kim, H., Kim, Y., and Lee, J. (2013). Liposomal formulations for enhanced lymphatic drug delivery. *Asian J. Pharm. Sci.* 8: 96–103.

60 Sha, X., Yan, G., Wu, Y. et al. (2005). Effect of selfmicroemulsifying drug delivery systems containing Labrasol on tight junctions in Caco-2 cells. *Eur. J. Pharm. Sci.* 24: 477–486.

61 Xu, X.F., Xu, P.P., Ma, C.W. et al. (2013). Gut microbiota, host health, and polysaccharides. *Biotechnol. Adv.* 31: 318–337.

62 Aqil, F., Munagala, R., Jeyabalan, J., and Vadhanam, M.V. (2013). Bioavailability of phytochemicals and its enhancement by drug delivery systems. *Cancer Lett.* 334: 133–141.

63 Kumari, A., Yadav, S.K., and Yadav, S.C. (2010). Biodegradable polymeric nanoparticles based drug delivery systems. *Colloids Surf., B* 75: 1–18.

64 Huang, E.J. and Reichardt, L.F. (2001). Neurotrophins: roles in neuronal development and function. *Annu. Rev. Neurosci.* 24: 677–736.

65 Mignani, S., El Kazzouli, S., Bousmina, M., and Majoral, J.P. (2013). Expand classical drug administration ways by emerging routes using dendrimer drug delivery systems: a concise overview. *Adv. Drug Delivery Rev.* 65: 1316–1330.

66 Lounnas, V., Ritschel, T., Kelder, J. et al. (2013). Current progress in structure-based rational drug design marks a new mindset in drug discovery. *Comput. Struct. Biotechnol. J.* 5: e201302011.

67 Mavromoustakos, T., Durdagi, S., Koukoulitsa, C. et al. (2011). Strategies in the rational drug design. *Curr. Med. Chem.* 18: 2517–2530.

68 Wong, P.T. and Choi, S.K. (2015). Mechanisms of drug release in nanotherapeutic delivery systems. *Chem. Rev.* 115: 3388–3432.

69 Prachayasittikul, V., Worachartcheewan, A., Shoombuatong, W. et al. (2015). Computer-aided drug design of bioactive natural products. *Curr. Top. Med. Chem.* 15: 1780–1800.

70 Chen, G., Roy, I., Yang, C., and Prasad, P.N. (2016). Nanochemistry and nanomedicine for nanoparticle-based diagnostics and therapy. *Chem. Rev.* 116: 2826–2885.

71 Pelaz, B., Alexiou, C., Alvarez-Puebla, R.A. et al. (2017). Diverse applications of nanomedicine. *ACS Nano* 11: 2313–2381.

72 Mattos, B.D., Rojas, O.J., and Magalhaes, W.L.E. (2017). Biogenic silica nanoparticles loaded with neem bark extract as green, slow-release biocide. *J. Cleaner Prod.* 142: 4206–4213.

73 Kinnear, C., Moore, T.L., Rodriguez-Lorenzo, L. et al. (2017). Form follows function: nanoparticle shape and its implications for nanomedicine. *Chem. Rev.* 117: 11476–11521.

74 Sethi, M., Sukumar, R., Karve, S. et al. (2014). Effect of drug release kinetics on nanoparticle therapeutic efficacy and toxicity. *Nanoscale* 6: 2321–2327.

75 Mattos, B.D., Tardy, B.L., Magalhaes, W.L.E., and Rojas, O.J. (2017). Controlled release for crop and wood protection: recent progress toward sustainable and safe nanostructured biocidal systems. *J. Controlled Release* 262: 139–150.

76 Siepmann, F., Herrmann, S., Winter, G., and Siepmann, J. (2008). A novel mathematical model quantifying drug release from lipid implants. *J. Controlled Release* 128: 233–240.

77 Patra, J.K., Das, G., Fraceto, L.F. et al. (2018). Nano based drug delivery systems: recent developments and future prospects. *J. Nanobiotechnol.* 16: 71. https://doi.org/10.1186/s12951-018-0392-8.

78 Lee, J.H. and Yeo, Y. (2015). Controlled drug release from pharmaceutical nanocarriers. *Chem. Eng. Sci.* 125: 75–84.

79 Kamaly, N., Yameen, B., Wu, J., and Farokhzad, O.C. (2016). Degradable controlled-release polymers and polymeric nanoparticles: mechanisms of controlling drug release. *Chem. Rev.* 116: 2602–2663.

80 Torchilin, V.P. (2012). Multifunctional nanocarriers. *Adv. Drug Delivery Rev.* 64: 302–315.

81 Pelaz, B., del Pino, P., Mafre, P. et al. (2015). Surface functionalization of nanoparticles with polyethylene glycol: effects on protein adsorption and cellular uptake. *ACS Nano* 9: 6996–7008.

82 Almalik, A., Benabdelkamel, H., Masood, A. et al. (2017). Hyaluronic acid coated chitosan nanoparticles reduced the immunogenicity of the formed protein corona. *Sci. Rep.* 7: 10542.

83 Martens, T.F., Remaut, K., Deschout, H. et al. (2015). Coating nanocarriers with hyaluronic acid facilitates intravitreal drug delivery for retinal gene therapy. *J. Controlled Release* 202: 83–92.

84 Kolhar, P., Anselmo, A.C., Gupta, V. et al. (2013). Using shape effects to target antibody-coated nanoparticles to lung and brain endothelium. *Proc. Natl. Acad. Sci. U.S.A.* 110: 10753–10758.

85 Jain, A. and Jain, S.K. (2015). Ligand-appended BBB-targeted nanocarriers (LABTNs). *Crit. Rev. Ther. Drug Carrier Syst.* 32: 149–180.

86 Shen, H.X., Shi, S.J., Zhang, Z.R. et al. (2015). Coating solid lipid nanoparticles with hyaluronic acid enhances antitumor activity against melanoma stem-like cells. *Theranostics* 5: 755–771.

87 Wang, T., Hou, J.H., Su, C. et al. (2017). Hyaluronic acid-coated chitosan nanoparticles induce ROS-mediated tumor cell apoptosis and enhance antitumor efficiency by targeted drug delivery via CD44. *J. Nanobiotechnol.* 15: 7.

88 Muro, S. (2012). Challenges in design and characterization of ligand-targeted drug delivery systems. *J. Controlled Release* 164: 125–137.

89 Kou, L., Sun, J., Zhai, Y., and He, Z. (2013). The endocytosis and intracellular fate of nanomedicines: implication for rational design. *Asian J. Pharm. Sci.* 8: 1–10.

90 Li, Z., Zhang, Y., Zhu, D. et al. (2017). Transporting carriers for intracellular targeting delivery via non-endocytic uptake pathways. *Drug Delivery* 24: 45–55.

91 Salatin, S. and Yari Khosroushahi, A. (2017). Overviews on the cellular uptake mechanism of polysaccharide colloidal nanoparticles. *J. Cell. Mol. Med.* 21: 1668–1686.

92 Anirudhan, T.S. and Nair, A.S. (2018). Temperature and ultrasound sensitive gatekeepers for the controlled release of chemotherapeutic drugs from mesoporous silica nanoparticles. *J. Mater. Chem. B* 6: 428–439.

93 Al-Ahmady, Z. and Kostarelos, K. (2016). Chemical components for the design of temperature-responsive vesicles as cancer therapeutics. *Chem. Rev.* 116: 3883–3918.

94 Bai, Y., Xie, F.Y., and Tian, W. (2018). Controlled self-assembly of thermo-responsive amphiphilic H-shaped polymer for adjustable drug release. *Chin. J. Polym. Sci.* 36: 406–416.

95 Zhang, Z., Zhang, D., Wei, L. et al. (2017). Temperature responsive fluorescent polymer nanoparticles (TRFNPs) for cellular imaging and controlled releasing of drug to living cells. *Colloids Surf., B* 159: 905–912.

96 Guo, Y., Zhang, Y., Ma, J. et al. (2017). Light/magnetic hyperthermia triggered drug released from multi-functional thermo-sensitive magnetoliposomes for precise cancer synergetic theranostics. *J. Controlled Release* 272: 145–158.

97 Hervault, A. and Thanh, N.T. (2014). Magnetic nanoparticle-based therapeutic agents for thermo-chemotherapy treatment of cancer. *Nanoscale* 6: 11553–11573.

98 Mathiyazhakan, M., Wiraja, C., and Xu, C.J. (2018). A concise review of gold nanoparticles-based photo-responsive liposomes for controlled drug delivery. *Nano-Micro Lett.* 10.

99 Xu, L., Qiu, L.Z., Sheng, Y. et al. (2018). Biodegradable pH-responsive hydrogels for controlled dual-drug release. *J. Mater. Chem. B* 6: 510–517.

100 Ma, G.L., Lin, W.F., Yuan, Z.F. et al. (2017). Development of ionic strength/pH/enzyme triple-responsive zwitterionic hydrogel of the mixed L-glutamic acid and L-lysine polypeptide for site-specific drug delivery. *J. Mater. Chem. B* 5: 935–943.

101 Grillo, R., Gallo, J., Stroppa, D.G. et al. (2016). Sub-micrometer magnetic nanocomposites: insights into the effect of magnetic nanoparticles interactions on the optimization of SAR and MRI performance. *ACS Appl. Mater. Interfaces* 8: 25777–25787.

102 Alonso, J., Khurshid, H., Devkota, J. et al. (2016). Superparamagnetic nanoparticles encapsulated in lipid vesicles for advanced magnetic hyperthermia and biodetection. *J. Appl. Phys.* 119: 083904.

103 Ulbrich, K., Hola, K., Subr, V. et al. (2016). Targeted drug delivery with polymers and magnetic nanoparticles: covalent and noncovalent approaches, release control, and clinical studies. *Chem. Rev.* 116: 5338–5431.

104 Robinson, M. and Zhang, X. (2011). *The World Medicines Situation. Traditional Medicines: Global Situation, Issues and Challenges*, 1–12. Geneva: World Health Organization.

105 Atanasov, A.G., Waltenberger, B., Pferschy-Wenzig, E.-M. et al. (2015). Discovery and resupply of pharmacologically active plant-derived natural products: a review. *Biotechnol. Adv.* 33: 1582–1614.

106 Namdari, M., Eatemadi, A., Soleimaninejad, M., and Hammed, A.T. (2017). A brief review on the application of nanoparticle enclosed herbal medicine for the treatment of infective endocarditis. *Biomed. Pharmacother.* 87: 321–331.

107 Kinghorn, A.D., Pan, L., Fletcher, J.N., and Chai, H. (2011). The relevance of higher plants in lead compound discovery programs. *J. Nat. Prod.* 74: 1539–1555.

108 Yuan, H., Ma, Q., Ye, L., and Piao, G. (2016). The traditional medicine and modern medicine from natural products. *Molecules* 21: 559.

109 Liu, G., Li, W., Qin, X., and Zhong, Q. (2020). Pickering emulsions stabilized by amphiphilic anisotropic nanofibrils of glycated whey proteins. *Food Hydrocolloids* 101: 105503.

110 Patra, J.K., Das, G., and Baek, K.-H. (2016). Towards a greener environment: synthesis and applications of green nanoparticles. *Pak. J. Agric. Sci.* 53: 59–79.

111 Duncan, R. and Gaspar, R. (2011). Nanomedicine(s) under the microscope. *Mol. Pharm.* 8: 2101–2141.

112 Ramana, K.V., Singhal, S.S., and Reddy, A.B. (2014). Therapeutic potential of natural pharmacological agents in the treatment of human diseases. *BioMed Res. Int.* 2014: 573452.

113 Guo, W. (2013). Green technology for nanoparticles in biomedical applications. In: *Green Biosynthesis of Nanoparticles: Mechanisms and Applications* (eds. M. Rai and C. Posten), 1–12. Wallington: CABI.

114 Wicki, A., Witzigmann, D., Balasubramanian, V., and Huwyler, J. (2015). Nanomedicine in cancer therapy: challenges, opportunities, and clinical applications. *J. Controlled Release* 200: 138–157.

115 Brigger, I., Dubernet, C., and Couvreur, P. (2002). Nanoparticles in cancer therapy and diagnosis. *Adv. Drug Delivery Rev.* 54: 631–651.

116 Ambesh, P., Campia, U., Obiagwu, C. et al. (2017). Nanomedicine in coronary artery disease. *Indian Heart J.* 69: 244–251.

117 Grazu, V., Moros, M., and Sánchez-Espinel, C. (2012). Nanocarriers as nanomedicines: design concepts and recent advances. In: *Frontiers of Nanoscience*, vol. 4 (eds. R.E. Palmer and R. Johnston), 337–440. New York: Elsevier.

118 Rizzo, L.Y., Theek, B., Storm, G. et al. (2013). Recent progress in nanomedicine: therapeutic, diagnostic and theranostic applications. *Curr. Opin. Biotechnol.* 24: 1159–1166.

119 Devasena, T. (2017). Diagnostic and therapeutic nanomaterials. In: *Therapeutic and Diagnostic Nanomaterials* (ed. T. Devasena), 1–13. New York: Springer.

120 Ventola, C.L. (2017). Progress in nanomedicine: approved and investigational nanodrugs. *Pharm. Ther.* 42: 742.

121 Kumar, A., Chen, F., Mozhi, A. et al. (2013). Innovative pharmaceutical development based on unique properties of nanoscale delivery formulation. *Nanoscale* 5: 8307–8325.

122 Boroumand Moghaddam, A., Namvar, F., Moniri, M. et al. (2015). Nanoparticles biosynthesized by fungi and yeast: a review of their preparation, properties, and medical applications. *Molecules* 20: 16540–16565.

123 Metz, K.M., Sanders, S.E., Pender, J.P. et al. (2015). Green synthesis of metal nanoparticles via natural extracts: the biogenic nanoparticle corona and its effects on reactivity. *ACS Sustainable Chem. Eng.* 3: 1610–1617.

124 Iravani, S. (2014). Bacteria in nanoparticle synthesis: current status and future prospects. *Int. Sch Res. Notices* 2014: 359316.

125 Mittal, A.K., Chisti, Y., and Banerjee, U.C. (2013). Synthesis of metallic nanoparticles using plant extracts. *Biotechnol. Adv.* 31: 346–356.

126 Khan, H.A., Sakharkar, M.K., Nayak, A. et al. (2018). 14 - Nanoparticles for biomedical applications: an overview. In: *Nanobiomaterials* (ed. R. Narayan), 357–384. Cambridge: Woodhead Publishing.

127 Aravamudhan, A., Ramos, D.M., Nada, A.A., and Kumbar, S.G. (2014). Natural polymers: polysaccharides and their derivatives for biomedical applications. In: *Natural and Synthetic Biomedical Polymers* (eds. S. Kumbar, C. Laurencin and M. Deng), 67–89. New York: Elsevier.

128 Luo, Y., Teng, Z., Wang, T.T.Y., and Wang, Q. (2013). Cellular uptake and transport of zein nanoparticles: effects of sodium caseinate. *J. Agric. Food Chem.* 61 (31): 7621–7629.

129 Pajardi, G., Rapisarda, V., Somalvico, F. et al. (2016). Skin substitutes based on allogenic fibroblasts or keratinocytes for chronic wounds not responding to conventional therapy: a retrospective observational study. *Int. Wound J.* 13: 44–52.

130 Rahimi, G., Alizadeh, F., and Khodavandi, A. (2016). Mycosynthesis of silver nanoparticles from *Candida albicans* and its antibacterial activity against *Escherichia coli* and *Staphylococcus aureus*. *Trop. J. Pharm. Res.* 15: 371–375.

131 Ali, M., Kim, B., Belfeld, K.D. et al. (2015). Inhibition of *Phytophthora parasitica* and *P. capsici* by silver nanoparticles synthesized using aqueous extract of *Artemisia absinthium*. *Phytopathology* 105: 1183–1190.

132 Malapermal, V., Botha, I., Krishna, S.B.N., and Mbatha, J.N. (2017). Enhancing antidiabetic and antimicrobial performance of *Ocimum basilicum*, and *Ocimum sanctum* (L.) using silver nanoparticles. *Saudi J. Biol. Sci.* 24: 1294–1305.

133 Sankar, R., Karthik, A., Prabu, A. et al. (2013). Origanum vulgare mediated biosynthesis of silver nanoparticles for its antibacterial and anticancer activity. *Colloids Surf., B* 108: 80–84.

134 Patra, J.K., Ali, M.S., Oh, I.-G., and Baek, K.-H. (2017). Proteasome inhibitory, antioxidant, and synergistic antibacterial and anticandidal activity of green biosynthesized magnetic Fe_3O_4 nanoparticles using the aqueous extract of corn (*Zea mays* L.) ear leaves. *Artif. Cells Nanomed. Biotechnol.* 45: 349–356.

135 Patra, J.K. and Baek, K.-H. (2017). Antibacterial activity and synergistic antibacterial potential of biosynthesized silver nanoparticles against foodborne pathogenic bacteria along with its anticandidal and antioxidant effects. *Front. Microbiol.* 8: 167.

136 Patra, J.K., Kwon, Y., and Baek, K.-H. (2016). Green biosynthesis of gold nanoparticles by onion peel extract: synthesis, characterization and biological activities. *Adv. Powder Technol.* 27: 2204–2213.

137 Patra, J.K. and Baek, K.-H. (2016). Biosynthesis of silver nanoparticles using aqueous extract of silky hairs of corn and investigation of its antibacterial and anticandidal synergistic activity and antioxidant potential. *IET Nanobiotechnol.* 10: 326–333.

138 Patra, J.K. and Baek, K.-H. (2016). Comparative study of proteasome inhibitory, synergistic antibacterial, synergistic anticandidal, and antioxidant activities of gold nanoparticles biosynthesized using fruit waste materials. *Int. J. Nanomed.* 11: 4691.

139 Patra, J.K. and Baek, K.-H. (2016). Green synthesis of silver chloride nanoparticles using *Prunus persica* L. outer peel extract and investigation of antibacterial, anticandidal, antioxidant potential. *Green Chem. Lett. Rev.* 9: 132–142.

140 Patra, J.K., Das, G., and Baek, K.-H. (2016). Phyto-mediated biosynthesis of silver nanoparticles using the rind extract of watermelon (*Citrullus lanatus*) under photo-catalyzed condition and investigation of its antibacterial, anticandidal and antioxidant efficacy. *J. Photochem. Photobiol., B* 161: 200–210.

141 Dias, D.A., Urban, S., and Roessner, U. (2012). A historical overview of natural products in drug discovery. *Metabolites* 2: 303–336.

142 Gupta, U., Sharma, S., Khan, I. et al. (2017). Enhanced apoptotic and anticancer potential of paclitaxel loaded biodegradable nanoparticles based on chitosan. *Int. J. Biol. Macromol.* 98: 810–819.

143 Panyam, J. and Labhasetwar, V. (2004). Sustained cytoplasmic delivery of drugs with intracellular receptors using biodegradable nanoparticles. *Mol. Pharm.* 1: 77–84.

144 Chang, C.-H., Huang, W.-Y., Lai, C.-H. et al. (2011). Development of novel nanoparticles shelled with heparin for berberine delivery to treat *Helicobacter pylori*. *Acta Biomater.* 7: 593–603.

145 Aldawsari, H.M. and Hosny, K.M. (2018). Solid lipid nanoparticles of vancomycin loaded with ellagic acid as a tool for overcoming nephrotoxic side effects: preparation, characterization, and nephrotoxicity evaluation. *J. Drug Delivery Sci. Technol.* 45: 76–80.

146 Dian, L., Yu, E., Chen, X. et al. (2014). Enhancing oral bioavailability of quercetin using novel soluplus polymeric micelles. *Nanoscale Res. Lett.* 9: 684.

147 Spillmann, C.M., Naciri, J., Algar, W.R. et al. (2014). Multifunctional liquid crystal nanoparticles for intracellular fluorescent imaging and drug delivery. *ACS Nano* 8: 6986–6997.

148 Lampe, K.J., Kem, D.S., Mahoney, M.J., and Bjugstad, K.B. (2011). The administration of BDNF and GDNF to the brain via PLGA microparticles patterned within a degradable PEG-based hydrogel: protein distribution and the glial response. *J. Biomed. Mater. Res.* 96: 595–607.

149 Purama, R.K., Goswami, P., Khan, A.T., and Goyal, A. (2009). Structural analysis and properties of dextran produced by *Leuconostoc mesenteroides* NRRL B-640. *Carbohydr. Polym.* 76: 30–35.

150 Huang, X. and Brazel, C.S. (2001). On the importance and mechanisms of burst release in matrix-controlled drug delivery systems. *J. Controlled Release* 73: 121–136.

151 Agarwal, A., Gupta, U., Asthana, A., and Jain, N.K. (2009). Dextran conjugated dendritic nanoconstructs as potential vectors for anti-cancer agent. *Biomaterials* 30: 3588–3596.

152 Barenholz, Y.C. (2012). Doxil® – the first FDA-approved nano-drug: lessons learned. *J. Controlled Release* 160: 117–134.

153 Maeng, J.H., Lee, D.-H., Jung, K.H. et al. (2010). Multifunctional doxorubicin loaded superparamagnetic iron oxide nanoparticles for chemotherapy and magnetic resonance imaging in liver cancer. *Biomaterials* 31: 4995–5006.

154 Bonechi, C., Martini, S., Ciani, L. et al. (2012). Using liposomes as carriers for polyphenolic compounds: the case of trans-resveratrol. *PLoS One* 7: e41438.

155 Noorafshan, A. and Ashkani-Esfahani, S. (2013). A review of therapeutic effects of curcumin. *Curr. Pharm. Des.* 19: 2032–2046.

156 Feng, T., Wei, Y., Lee, R.J., and Zhao, L. (2017). Liposomal curcumin and its application in cancer. *Int. J. Nanomed.* 12: 6027.

157 Cheng, C., Peng, S., Li, Z. et al. (2017). Improved bioavailability of curcumin in liposomes prepared using a pH-driven, organic solvent-free, easily scalable process. *RSC Adv.* 7: 25978–25986.

158 Carroll, R.J., Thompson, M.P., and Nutting, G.C. (1968). Glutaraldehyde fixation of casein micelles for electron microscopy. *J. Dairy Sci.* 51 (12): 1903–1908.

159 Wusigale, Liang, Li., and Luo, Y. (2020). Casein and pectin: structures, interactions, and applications. *Trends Food Sci. Technol.* 97: 391–403.

160 Huppertz, T., Gazi, I., Luyten, H. et al. (2017). Hydration of casein micelles and caseinates: implications for casein micelle structure. *Int. Dairy J.* 74: 1–11.

161 Elzoghby, A.O., Abo El-Fotoh, W.S., and Elgindy, N.A. (2011). Casein-based formulations as promising controlled release drug delivery systems. *J. Controlled Release* 153 (3): 206–216.

162 Semo, E., Kesselman, E., Danino, D., and Livney, Y.D. (2007). Casein micelle as a natural nano-capsular vehicle for nutraceuticals. *Food Hydrocolloids* 21 (5): 936–942.

163 Luo, Y., Wang, Q., and Zhang, Y. (2020). Biopolymer-based nanotechnology approaches to deliver bioactive compounds for food applications: a perspective on the past, present, and future. *J. Agric. Food Chem.* https://doi.org/10.1021/acs.jafc.0c00277.

164 Pan, K., Luo, Y., Gan, Y. et al. (2014). pH-driven encapsulation of curcumin in self-assembled casein nanoparticles for enhanced dispersibility and bioactivity. *Soft Matter* 10 (35): 6820–6830.

165 Wang, L. and Zhang, Y. (2017). Eugenol nanoemulsion stabilized with zein and sodium caseinate by self-assembly. *J. Agric. Food Chem.* 65 (14): 2990–2998.

166 Chen, Y., Xue, J., Wusigale et al. (2020). Carboxymethylation of phytoglycogen and its interactions with caseinate for the preparation of nanocomplex. *Food Hydrocolloids* 100: 105390.

167 Rodriguez, N.J., Hu, Q., and Luo, Y. (2019). Oxidized dextran as a macromolecular crosslinker stabilizes the zein/caseinate nanocomplex for the potential oral delivery of curcumin. *Molecules* 24 (22): 4061.

168 Xue, J., Wang, T., Hu, Q. et al. (2018). Insight into natural biopolymer-emulsified solid lipid nanoparticles for encapsulation of curcumin: effect of loading methods. *Food Hydrocolloids* 79: 110–116.

169 Zhang, Y., Yao, F., Liu, J. et al. (2018). Synthesis and characterization of alkylated caseinate, and its structure-curcumin loading property relationship in water. *Food Chem.* 244: 246–253.

170 Luo, Y. and Wang, Q. (2014). Zein-based micro- and nano-particles for drug and nutrient delivery: a review. *J. Appl. Polym. Sci.* 131 (16): 40696.

171 Wang, Q., Xian, W., Li, S. et al. (2008). Topography and biocompatibility of patterned hydrophobic/hydrophilic zein layers. *Acta Biomater.* 4 (4): 844–851.

172 Zhong, Q. and Jin, M. (2009). Zein nanoparticles produced by liquid–liquid dispersion. *Food Hydrocolloids* 23 (8): 2380–2387.

173 Wang, Y. and Padua, G.W. (2012). Nanoscale characterization of zein self-assembly. *Langmuir* 28 (5): 2429–2435.

174 Hu, K. and McClements, D.J. (2014). Fabrication of surfactant-stabilized zein nanoparticles: a pH modulated antisolvent precipitation method. *Food Res. Int.* 64: 329–335.

175 Patel, A.R., Bouwens, E.C.M., and Velikov, K.P. (2010). Sodium caseinate stabilized zein colloidal particles. *J. Agric. Food Chem.* 58 (23): 12497–12503.

176 Xu, H., Jiang, Q., Reddy, N., and Yang, Y. (2011). Hollow nanoparticles from zein for potential medical applications. *J. Mater. Chem.* 21 (45): 18227–18235.

177 Hu, S., Wang, T., Fernandez, M.L., and Luo, Y. (2016). Development of tannic acid cross-linked hollow zein nanoparticles as potential oral delivery vehicles for curcumin. *Food Hydrocolloids* 61: 821–831.

178 Zou, T. and Gu, L. (2013). TPGS emulsified zein nanoparticles enhanced oral bioavailability of daidzin: in vitro characteristics and in vivo performance. *Mol. Pharm.* 10 (5): 2062–2070.

179 Xu, H., Shen, L., Xu, L., and Yang, Y. (2015). Controlled delivery of hollow corn protein nanoparticles via non-toxic crosslinking: in vivo and drug loading study. *Biomed. Microdevices* 17 (1): 8.

180 Teng, Z., Luo, Y., and Wang, Q. (2012). Nanoparticles synthesized from soy protein: preparation, characterization, and application for nutraceutical encapsulation. *J. Agric. Food Chem.* 60 (10): 2712–2720.

181 Liu, L.-L., Liu, P.-Z., Li, X.-T. et al. (2019). Novel soy β-conglycinin core–shell nanoparticles as outstanding ecofriendly nanocarriers for curcumin. *J. Agric. Food Chem.* 67 (22): 6292–6301.

182 Yi, J., Lam, T.I., Yokoyama, W. et al. (2015). Beta-carotene encapsulated in food protein nanoparticles reduces peroxyl radical oxidation in Caco-2 cells. *Food Hydrocolloids* 43: 31–40.

183 Teng, Z., Luo, Y., and Wang, Q. (2013). Carboxymethyl chitosan–soy protein complex nanoparticles for the encapsulation and controlled release of vitamin D_3. *Food Chem.* 141 (1): 524–532.

184 Gao, Z.-M., Zhu, L.-P., Yang, X.-Q. et al. (2014). Soy lipophilic protein nanoparticles as a novel delivery vehicle for conjugated linoleic acid. *Food Funct.* 5 (6): 1286–1293.

185 Zhu, X.-F., Zheng, J., Liu, F. et al. (2018). Freeze-thaw stability of Pickering emulsions stabilized by soy protein nanoparticles. Influence of ionic strength before or after emulsification. *Food Hydrocolloids* 74: 37–45.

186 Teng, Z., Xu, R., and Wang, Q. (2015). Beta-lactoglobulin-based encapsulating systems as emerging bioavailability enhancers for nutraceuticals: a review. *RSC Adv.* 5 (44): 35138–35154.

187 Joye, I.J., Nelis, V.A., and McClements, D.J. (2015). Gliadin-based nanoparticles: fabrication and stability of food-grade colloidal delivery systems. *Food Hydrocolloids* 44: 86–93.

188 Sirisha, V.L. and D'Souza, J.S. (2016). Polysaccharide-based nanoparticles as drug delivery systems: principles and applications. *Adv. Drug Delivery Rev.* 60: 1650–1662. https://doi.org/10.1201/9781315372303-32.

189 Hu, Q. and Luo, Y. (2018). Recent advances of polysaccharide-based nanoparticles for oral insulin delivery. *Int. J. Biol. Macromol.* 120: 775–782.

190 Liang, J., Yan, H., Puligundla, P. et al. (2017). Applications of chitosan nanoparticles to enhance absorption and bioavailability of tea polyphenols: a review. *Food Hydrocolloids* 69: 286–292.

191 Puligundla, P., Mok, C., Ko, S. et al. (2017). Nanotechnological approaches to enhance the bioavailability and therapeutic efficacy of green tea polyphenols. *J. Funct. Foods* 34: 139–151.

192 Grant, G.T., Morris, E.R., Rees, D.A. et al. (1973). Biological interactions between polysaccharides and divalent cations: the egg-box model. *FEBS Lett.* 32 (1): 195–198.

193 Lee, K.Y. and Mooney, D.J. (2012). Alginate: properties and biomedical applications. *Prog. Polym. Sci.* 37 (1): 106–126.

194 Đordević, V., Balanč, B., Belščak-Cvitanović, A. et al. (2015). Trends in encapsulation technologies for delivery of food bioactive compounds. *Food Eng. Rev.* 7 (4): 452–490.

195 Istenič, K., Balanč, B.D., Djordjević, V.B. et al. (2015). Encapsulation of resveratrol into Ca-alginate submicron particles. *J. Food Eng.* 167: 196–203.

196 Bagheri, L., Madadlou, A., Yarmand, M., and Mousavi, M.E. (2014). Spray-dried alginate microparticles carrying caffeine-loaded and potentially bioactive nanoparticles. *Food Res. Int.* 62: 1113–1119.

197 Li, Q., Liu, C.-G., and Yu, Y. (2015). Separation of monodisperse alginate nanoparticles and effect of particle size on transport of vitamin E. *Carbohydr. Polym.* 124: 274–279.

198 Klemm, D., Kramer, F., Moritz, S. et al. (2011). Nanocelluloses: a new family of nature-based materials. *Angew. Chem. Int. Ed.* 50 (24): 5438–5466.

199 Gómez H., C., Serpa, A., Velásquez-Cock, J. et al. (2016). Vegetable nanocellulose in food science: a review. *Food Hydrocolloids* 57: 178–186.

200 Ullah, H., Santos, H.A., and Khan, T. (2016). Applications of bacterial cellulose in food, cosmetics and drug delivery. *Cellulose* 23 (4): 2291–2314.

201 Chen, H. and Yao, Y. (2017). Phytoglycogen to increase lutein solubility and its permeation through Caco-2 monolayer. *Food Res. Int.* 97: 258–264.

202 Xue, J., Inzero, J., Hu, Q. et al. (2019). Development of easy, simple and low-cost preparation of highly purified phytoglycogen nanoparticles from corn. *Food Hydrocolloids* 95: 256–261.

203 Ye, F., Miao, M., Cui, S.W. et al. (2018). Characterisations of oil-in-water Pickering emulsion stabilized hydrophobic phytoglycogen nanoparticles. *Food Hydrocolloids* 76: 78–87.

204 Chen, H. and Zhong, Q. (2014). Processes improving the dispersibility of spray-dried zein nanoparticles using sodium caseinate. *Food Hydrocolloids* 35: 358–366.

205 Xue, J., Zhang, Y., Huang, G. et al. (2018). Zein-caseinate composite nanoparticles for bioactive delivery using curcumin as a probe compound. *Food Hydrocolloids* 83: 25–35.

206 Wang, L. and Zhang, Y. (2019). Heat-induced self-assembly of zein nanoparticles: fabrication, stabilization and potential application as oral drug delivery. *Food Hydrocolloids* 90: 403–412.

207 Chang, C., Wang, T., Hu, Q. et al. (2017). Pectin coating improves physicochemical properties of caseinate/zein nanoparticles as oral delivery vehicles for curcumin. *Food Hydrocolloids* 70: 143–151.

208 Hu, Q., Hu, S., Fleming, E. et al. (2020). Chitosan-caseinate-dextran ternary complex nanoparticles for potential oral delivery of astaxanthin with significantly improved bioactivity. *Int. J. Biol. Macromol.* 151: 747–756.

209 Wang, T., Xue, J., Hu, Q. et al. (2017). Synthetic surfactant- and cross-linker-free preparation of highly stable lipid-polymer hybrid nanoparticles as potential oral delivery vehicles. *Sci. Rep.* 7 (1): 2750.

210 Wang, T., Hu, Q., Lee, J.-Y., and Luo, Y. (2018). Solid lipid–polymer hybrid nanoparticles by in situ conjugation for oral delivery of astaxanthin. *J. Agric. Food Chem.* 66 (36): 9473–9480.

211 Chu, P.Y., Tsai, S.C., Ko, H.Y. et al. (2019). Co-delivery of natural compounds with a dual-targeted nanoparticle delivery system for improving synergistic therapy in an orthotopic tumor model. *ACS Appl. Mater. Interfaces* 11: 23880–23892.

212 Pandit, A.P., Joshi, S.R., Dalal, P.S., and Patole, V.C. (2019). Curcumin as a permeability enhancer enhanced the antihyperlipidemic activity of dietary green tea extract. *BMC Complement. Altern. Med.* 19 https://doi.org/10.1186/s12906-019-2545-1.

213 Yan, X., Zhang, X., McClements, D.J. et al. (2019). Co-encapsulation of epigallocatechin gallate (EGCG) and curcumin by two proteins-based nanoparticles: role of EGCG. *J. Agric. Food Chem.* 67: 13228–13236.

214 Chen, S., Han, Y., Huang, J. et al. (2019). Fabrication and characterization of layer-by-layer composite nanoparticles based on zein and hyaluronic acid for codelivery of curcumin and quercetagetin. *ACS Appl. Mater. Interfaces* 11: 16922–16933.

215 Chen, S., McClements, D.J., Jian, L. et al. (2019). Core–shell biopolymer nanoparticles for co-delivery of curcumin and piperine: sequential electrostatic deposition of hyaluronic acid and chitosan shells on the zein core. *ACS Appl. Mater. Interfaces* 11: 38103–38115.

216 Chen, W.J., Zou, M.M., Ma, X.B. et al. (2019). Co-encapsulation of EGCG and quercetin in liposomes for optimum antioxidant activity. *J. Food Sci.* 84: 111–120.

217 Das, A., Konyak, P.M., Das, A. et al. (2019). Physicochemical characterization of dual action liposomal formulations: anticancer and antimicrobial. *Heliyon* 5 https://doi.org/10.1016/j.heliyon.2019.e02372.

218 Fomina, N., Sankaranarayanan, J., and Almutairi, A. (2012). Photochemical mechanisms of light-triggered release from nanocarriers. *Adv. Drug Delivery Rev.* 64: 1005–1020.

219 Alvarez-Lorenzo, C., Bromberg, L., and Concheiro, A. (2009). Light-sensitive intelligent drug delivery systems. *Photochem. Photobiol.* 85: 848–860.

220 Bisby, R.H., Mead, C., and Morgan, C.G. (2000). Wavelength-programmed solute release from photosensitive liposomes. *Biochem. Biophys. Res. Commun.* 276: 169–173.

221 Morgan, C.G., Yianni, Y.P., Sandhu, S.S., and Mitchell, A.C. (1995). Liposome fusion and lipid exchange on ultraviolet irradiation of liposomes containing a photochromic phospholipid. *Photochem. Photobiol.* 62: 24–29.

222 Bareford, L.M. and Swaan, P.W. (2007). Endocytic mechanisms for targeted drug delivery. *Adv. Drug Delivery Rev.* 59: 748–758.

223 Sha, X.Y., Guo, J., Chen, Y.Z., and Fang, X.L. (2012). Effect of phospholipid composition on pharmacokinetics and biodistribution of epirubicin liposomes. *J. Liposome Res.* 22: 80–88.

224 Angelova, A., Angelov, B., Drechsler, M., and Lesieur, S. (2013). Neurotrophin delivery using nanotechnology. *Drug Discovery Today* 18: 1263–1271.

225 Agrahari, V., Agrahari, V., Hung, W.T. et al. (2016). Composite nanoformulation therapeutics for long-term ocular delivery of macromolecules. *Mol. Pharm.* 13: 2912–2922.

226 Amin, F.U., Shah, S.A., Badshah, H. et al. (2017). Anthocyanins encapsulated by PLGA@PEG nanoparticles potentially improved its free radical scavenging capabilities via p38/JNK pathway against $A\beta_{1-42}$-induced oxidative stress. *J. Nanobiotechnol.* 15: 12–27.

227 Yamamoto, H., Tahara, K., and Kawashima, Y. (2012). Nanomedical system for nucleic acid drugs created with the biodegradable nanoparticle platform. *J. Microencapsulation* 29: 54–62.

228 Maurer-Jones, M.A., Bantz, K.C., Love, S.A. et al. (2009). Toxicity of therapeutic nanoparticles. *Nanomedicine* 4: 219–241.

229 Vega-Villa, K.R., Takemoto, J.K., Yáñez, J.A. et al. (2008). Clinical toxicities of nanocarrier systems. *Adv. Drug Delivery Rev.* 60: 929–938.

230 Curtis, J., Greenberg, M., Kester, J. et al. (2006). Nanotechnology and nanotoxicology: a primer for clinicians. *Toxicol. Sci.* 25: 245–260.

231 Lanone, S. and Boczkowski, J. (2006). Biomedical applications and potential health risks of nanomaterials: molecular mechanisms. *Curr. Mol. Med.* 6: 651–663.

232 Moghimi, S.M., Hunter, A.C., and Murray, J.C. (2005). Nanomedicine: current status and future prospects. *FASEB J.* 19: 311–330.

Index

Nano-Technological Intervention in Agricultural Productivity, First Edition.
Javid A. Parray, Mohammad Yaseen Mir and Nowsheen Shameem.